CLIMATE CHANGE AND ITS CAUSES, EFFECTS AND PREDICTION

CLIMATE CHANGE'S EFFECT ON INSURERS

EXPOSURES, RISKS, AND PREPARATION

CLIMATE CHANGE AND ITS CAUSES, EFFECTS AND PREDICTION

Additional books in this series can be found on Nova's website
under the Series tab.

Additional e-books in this series can be found on Nova's website
under the e-book tab.

CLIMATE CHANGE AND ITS CAUSES, EFFECTS AND PREDICTION

CLIMATE CHANGE'S EFFECT ON INSURERS

EXPOSURES, RISKS, AND PREPARATION

NOAH LYNCH
EDITOR

nova publishers
New York

Library of Congress Cataloging-in-Publication Data

ISBN: 978-1-63482-518-4

Published by Nova Science Publishers, Inc. † New York

CONTENTS

Preface **vii**

Chapter 1 Climate Change: Better Management of Exposure to Potential Future Losses Is Needed for Federal Flood and Crop Insurance **1**
United States Government Accountability Office

Chapter 2 Climate Change: Financial Risks to Federal and Private Insurers in Coming Decades Are Potentially Significant **41**
United States Government Accountability Office

Index **101**

PREFACE

This book examines how federal and private exposure to losses has changed since GAO's 2007 report on the subject, and what is known about how climate change may affect insured and uninsured losses; how public insurers are preparing for climate change, and any challenges they face; and how private insurers are preparing for climate change and any challenges they face.

Chapter 1 – The May 2014 National Climate Assessment indicates that the frequency and/or severity of many weather and climate extremes may increase with climate change. Public and private property insurers can bear a large portion of the financial impact of such weather-related losses. In the public sector, federal insurance includes NFIP, managed by FEMA, and the federal crop insurance program, managed by RMA.

GAO was asked to review climate change's effect on insurers. This report examines (1) how federal and private exposure to losses has changed since GAO's 2007 report on the subject, and what is known about how climate change may affect insured and uninsured losses; (2) how public insurers are preparing for climate change, and any challenges they face; and (3) how private insurers are preparing for climate change and any challenges they face. GAO reviewed 20 studies, examined federal and private sector data on exposure to losses from 2000 to 2013, reviewed agency documents, and interviewed agency officials and a nonprobability sample of eight insurers and reinsurers.

Chapter 2 – Weather-related events have cost the nation billions of dollars in damages over the past decade. Many of these losses are borne by private insurers and by two federal insurance programs—the National Flood Insurance Program (NFIP), which insures properties against flooding, and the Federal

Crop Insurance Corporation (FCIC), which insures crops against drought or other weather disasters.

GAO was asked to (1) describe how climate change may affect future weather-related losses, (2) determine past insured weather-related losses, and (3) determine what major private insurers and federal insurers are doing to prepare for potential increases in such losses. In response, among other things, GAO reviewed key scientific assessments; analyzed insured loss data; and contacted private insurers, NFIP, and FCIC.

In: Climate Change's Effect on Insurers
Editor: Noah Lynch
ISBN: 978-1-63482-518-4

Chapter 1

CLIMATE CHANGE: BETTER MANAGEMENT OF EXPOSURE TO POTENTIAL FUTURE LOSSES IS NEEDED FOR FEDERAL FLOOD AND CROP INSURANCE*

United States Government Accountability Office

WHY GAO DID THIS STUDY

The May 2014 National Climate Assessment indicates that the frequency and/or severity of many weather and climate extremes may increase with climate change. Public and private property insurers can bear a large portion of the financial impact of such weather-related losses. In the public sector, federal insurance includes NFIP, managed by FEMA, and the federal crop insurance program, managed by RMA.

GAO was asked to review climate change's effect on insurers. This report examines (1) how federal and private exposure to losses has changed since GAO's 2007 report on the subject, and what is known about how climate change may affect insured and uninsured losses; (2) how public insurers are preparing for climate change, and any challenges they face; and (3) how private insurers are preparing for climate change and any challenges they face.

---

* This is an edited, reformatted and augmented version of a United States Government Accountability Office publication, No. GAO-15-28, dated October 2014.

GAO reviewed 20 studies, examined federal and private sector data on exposure to losses from 2000 to 2013, reviewed agency documents, and interviewed agency officials and a nonprobability sample of eight insurers and reinsurers.

What GAO Recommends

GAO recommends that FEMA and RMA take additional steps to encourage flood and crop insurance policyholders to adopt building and agricultural practices that reduce long-term risk and federal exposure to losses. FEMA agreed with GAO's recommendation, and RMA neither agreed nor disagreed with GAO's recommendation.

What GAO Found

Since GAO's 2007 report on flood and crop insurance, exposure growth in hazard-prone areas has increased losses, and climate change and related increases in extreme weather events may further increase such losses in coming decades. Scientific and industry studies GAO reviewed generally found that increasing growth and property values in hazard-prone areas have increased losses to date and that climate change may compound this effect. From 2007 through 2013, data from the Federal Emergency Management Agency (FEMA) and the Risk Management Agency (RMA) show that exposure to potential losses for insured property grew from \$1.3 trillion to \$1.4 trillion (8 percent). According to industry data, private sector exposure to such loss grew from \$60.7 trillion to \$66.5 trillion (10 percent) from 2007 through 2012. Federal exposure to uninsured loss also increased by 46 percent over this period, based on a 2013 analysis by the Congressional Research Service. According to the studies GAO reviewed, climate change may substantially increase losses by 2040 and increase losses from about 50 to 100 percent by 2100.

FEMA and RMA have taken some steps to better understand and prepare for climate change's potential effects under the National Flood Insurance Program (NFIP) and the federal crop insurance program by, for example, commissioning climate change studies. However, both agencies face challenges that may limit their ability to minimize long-term federal exposure to climate change. For example, because of the short-term nature of insurance (i.e., contracts typically estimate and communicate risk of property losses for the 1-year term of a policy), FEMA and RMA face a challenge in encouraging policyholders to reduce their long-term exposure to climate change risks. Specifically, flood insurance policyholders who build to NFIP standards—which are based on current flood risk and not on long-term risks—may unintentionally increase their vulnerability to climate change as sea-level rises. Also, while federal law prohibits crop insurance from covering losses due to a farmers' failure to follow good farming practices, such as various irrigation methods, some of these practices maintain short-term production but may inadvertently increase the vulnerability of agriculture to climate change through increased erosion and inefficient water use. A recent executive order directed federal agencies to reform policies that may, perhaps unintentionally, increase the vulnerability of economic sectors or communities to climate change. Without encouraging NFIP and crop insurance policyholders to adopt building and agricultural practices that reduce long-term risk, FEMA and RMA may send policyholders signals that unintentionally encourage their vulnerability to climate change, counter to the direction of the executive order, which could exacerbate federal exposure to losses.

Many private insurers and reinsurers have taken steps since 2007 to better understand and prepare for climate change effects and related challenges, including participating in industry climate change surveys, and issuing reports that identify and assess climate change risks and trends in weather-related losses. According to industry officials, they can manage their exposure to climate change and related challenges through risk-based premiums, reinsurance, and other practices, although estimating weather-related risks still includes elements of uncertainty.

ABBREVIATIONS

ARS	Agricultural Research Service
Biggert-Waters Act	Biggert-Waters Flood Insurance Reform Act of 2012
DHS	Department of Homeland Security
DRF	Disaster Relief Fund
FAIR	Fair Access to Insurance Requirements
FEMA	Federal Emergency Management Agency
NAIC	National Association of Insurance Commissioners
NFIP	National Flood Insurance Program

NOAA	National Oceanic and Atmospheric Administration
NRC	National Research Council
PRISM	Parameter-elevation Regressions on Independent Slopes Model
RMA	Risk Management Agency
RMS	Risk Management Solutions
SFHA	special flood hazard area
USDA	U.S. Department of Agriculture

* * *

October 29, 2014

Congressional Requesters

The impacts and costs of certain weather disasters—such as floods, drought, and hurricanes—will increase in significance as previously rare events become more common and intense due to climate change, according to the United States Global Change Research Program's May 2014 National Climate Assessment.[1] While the magnitude and extent of future climate change effects are uncertain, climate change presents a significant financial risk to the federal government in many areas. As a result of these risks and the nation's fiscal condition, in February 2013, we added Limiting the Federal Government's Fiscal Exposure by Better Managing Climate Change Risks to our list of areas at high risk to fraud, waste, abuse, and mismanagement, or most in need of transformation.[2]

The uncertain and potentially large losses associated with weather-related events are among the biggest risks that property insurers face. The property and casualty segment of insurance, which spans both the public and private sectors, bears a large portion of the financial impact of weather-related losses. In the public sector, federal insurance programs include the National Flood Insurance Program (NFIP), administered by the Federal Emergency Management Agency (FEMA) within the U.S. Department of Homeland Security, which makes flood insurance available to residential property owners and businesses, and the federal crop insurance program, administered by the Risk Management Agency (RMA) within the U.S. Department of Agriculture (USDA), which protects farmers against financial losses caused by drought, floods, and other natural disasters, among other things. In addition, the federal government provides disaster assistance through FEMA and other federal agencies.[3] In certain states, insurance pools are administered by governmental entities or entities established pursuant to state law that provide coverage for losses caused by weather-related events. In the private sector, primary insurers provide insurance to individuals and businesses directly, and reinsurers provide insurance to primary insurers.

As of June 2014, FEMA's debt to the U.S. Treasury from flood insurance payments totaled about $24 billion—primarily to pay claims associated with Hurricane Katrina in 2005 and Hurricane Sandy in 2012—and FEMA had not repaid any principal on its loan since 2010. NFIP has been on our High-Risk List since March 2006 because of concerns about its long-term financial solvency and related operational issues.[4] While FEMA is implementing some changes to NFIP in response to recently enacted legislation, it is unclear to what extent these changes will improve the program's financial condition.[5] Additionally, in our 2013 High-Risk Update, we concluded that FEMA obligated over $80 billion in federal assistance for disasters declared during fiscal years 2004 through 2011.[6] As of June 2014, FEMA's Disaster Relief Fund had spent $3.8 billion for fiscal year 2014, with an unspent balance of over $11.3 billion. We have previously found that the federal government does not fully budget for these costs— in part because supplemental appropriations are largely used to fund them—obscuring the full cost of disaster relief commitments.[7]

Further, as we reported in 2014, the federal government's crop insurance costs have increased in recent years, with appropriations more than doubling from $3.4 billion in fiscal year 2001 to $7.6 billion in fiscal year 2012 as program expansion and rising crop prices have led to increasing subsidy values and higher claims payments.[8] In 2007, we found that climate change could increase federal insurers' financial risk and recommended that the Secretaries of Agriculture and Homeland Security analyze the potential long-term fiscal implications of climate change for crop insurance and flood insurance, respectively, and report their findings to Congress.[9] USDA and DHS agreed with these recommendations and in response issued reports containing such analyses.

You asked us to review the implications of climate change on public flood and crop insurance and private sector insurance. This report examines (1) how federal and private exposure to losses has changed since our 2007 report on flood and crop insurance and how climate change may affect insured and uninsured losses; (2) how public insurers are preparing for climate change and associated challenges, if any; and (3) how selected private insurers and reinsurers are preparing for climate change and associated challenges, if any.

To determine how exposure to losses has changed since our 2007 report on flood and crop insurance and how climate change may affect insured and uninsured losses, we reviewed scientific studies and other available literature on climate change and analyzed federal and private sector exposure data from 2007 through 2013. To understand longer-term trends in exposure, we also analyzed federal exposure data from 2000. We assessed the reliability of the exposure data and reviewed the methodological soundness of each study and determined that the data and studies were sufficiently reliable for describing trends in exposure and the potential effect of climate change on insured and uninsured losses.[10] To determine how public insurers are preparing for climate change, we reviewed FEMA and RMA documents and interviewed agency officials and a nonprobability sample of three state insurers that were identified by an industry forecaster as most at-risk of natural disasters, whether the state had an insurance pool administered by a governmental entity or entity established pursuant to state law, and whether the state had experienced an extreme weather event in the past decade.[11] Further, to determine how private insurers and reinsurers are preparing for climate change, we reviewed over 20 industry reports and other information from industry representatives we interviewed. We interviewed a nonprobability sample of eight insurers and reinsurers, representing over 25 percent of the U.S. property and casualty insurance market, 15 percent of the private crop insurance market, and over 30 percent of the U.S. reinsurance market. We selected both large and small insurers with relevant experience in flood or crop insurance, to collect a variety of perspectives. We also interviewed two catastrophe modeling firms, two reinsurance brokerage firms, and eight industry groups representing more than 1,000 large and small property and casualty insurers. Because we used a nonprobability sample to select interviewees, our results are not generalizable to the industry as a whole but provide illustrative examples. Appendix I presents a more detailed description of our objectives, scope, and methodology.

We conducted this performance audit from November 2013 to October 2014 in accordance with generally accepted government auditing standards. Those standards require that we plan and perform the audit to obtain sufficient, appropriate evidence to provide a reasonable basis for our findings and conclusions based on our audit objectives. We believe that the evidence obtained provides a reasonable basis for our findings and conclusions based on our audit objectives.

BACKGROUND

Insurance can spread risk over time, across geographical areas, and among industries and individuals. While private insurers assume some financial risk when they write policies, they employ various strategies to manage risk so that they limit potential financial exposures, earn profits, and build capital needed to pay claims. For example, insurers charge premiums for coverage and establish underwriting standards, such as refusing to insure customers who pose unacceptable levels of risk, or limiting coverage in particular geographic areas. Insurance companies may also purchase insurance from reinsurers to cover specific portions of their financial risk. Reinsurers use similar strategies to limit their risks, including charging premiums and establishing underwriting standards.[12] States regulate private insurance and may impose various restrictions on insurers' risk management practices, such as premium rate increases and coverage limitations, to protect consumers and to ensure insurer solvency. However, many of these risk management strategies are not available to NFIP, which is required to assume and retain all of the risk and to generally accept all insurance applicants, including those with potentially high-risk properties.

The uncertain and potentially large losses associated with weather-related events are among the biggest risks that property insurers face. Virtually anything that is insured—property, crops and livestock, business operations, or human life and health—is vulnerable to weather-related events. For insurers, remaining financially solvent generally involves estimating and setting rates that reflect insured risks; any unanticipated changes in the frequency or severity of weather-related events can have financial consequences for them. Recent scientific assessments have found that climate change has or will alter the frequency and/or severity of damaging weather-related events—such as droughts and floods, alter crop productivity, and threaten coastal communities as sea-level rises.[13]

Under certain circumstances, the private sector may consider some risks uninsurable. In other instances, the private sector may offer to insure a risk, but at rates that many property owners cannot afford. Without insurance, affected property owners must rely on their own resources or seek disaster assistance from local, state, and federal sources in the event of a loss. In situations where the private sector will not insure a particular type of risk, the public sector may create markets to ensure the availability of insurance. For example, several states have established Fair Access to Insurance Requirements (FAIR) and windstorm plans, that pool resources from insurers doing business in the state to make insurance available to property owners who either cannot obtain coverage in the private insurance market or cannot do so at an affordable rate.

Similarly, at the federal level, the NFIP and the federal crop insurance program were established to provide coverage where private markets did not exist, and partly to provide an alternative to disaster assistance. NFIP has three components: (1) the provision of flood insurance; (2) the requirement that participating communities adopt and enforce floodplain management regulations that are at least as stringent as FEMA's national minimum standards; and (3) the identification and mapping of floodplains, which helps to determine which insurance premiums and regulations apply to a particular property. Community participation in NFIP is voluntary. However, communities must join NFIP and adopt FEMA-approved building standards and floodplain management strategies in order for their residents to purchase flood insurance through the program.[14] In addition to meeting these federal standards, the regulations that each community enacts and implements must meet the minimum state requirements, which are established consistent with NFIP standards.[15] For example, communities must adopt at least the minimum standards for floodplain management regulations, including building requirements to reduce future flood damage, such as requiring new and substantially improved or substantially damaged structures in special flood hazard areas to be elevated to or above base flood elevation levels. Additionally, FEMA provides premium reduction incentives for policyholders within communities that take measures to mitigate flood risk beyond NFIP minimum requirements through the agency's program, the Community Rating System.[16] Under NFIP, the federal government assumes the liability for covered losses and sets rates and coverage limitations. While NFIP does not subsidize most policies, policyholders with certain types of insured properties pay subsidized premiums.[17]

For crop insurance, farmers participate voluntarily, but the federal government encourages participation by subsidizing their insurance premiums. USDA's RMA administers the federal crop insurance program, including issuing new insurance products and expanding existing products to new geographic regions. RMA administers the program in partnership with private insurance companies that share a percentage of the risk of loss or the opportunity for gain associated with each policy. Federal law prohibits crop insurance from covering losses due to a farmer's failure to follow good farming practices. Good farming practices are identified by agricultural experts in a given area and provide acceptable farming methods for crop insurance policyholders to use in producing yields consistent with historical production. Agricultural experts are individuals employed by USDA's Cooperative Extension System or the agricultural departments of universities, or other approved persons, whose research or occupation is related to the specific crop or practice for which expertise is sought.

The federal government's disaster relief and flood and crop insurance programs create fiscal exposure to weather-related events and climate change. For the purposes of this report, fiscal exposures are responsibilities, programs, and activities that may either obligate the federal government to future spending or create the expectation for future spending.[18] For example, the government's response to a weather-related event or series of events can strengthen expectations that the government will respond in the same way to similar events in the future.

Differences exist in the types of property that public insurers insure, as well as their time horizons. For example, federal flood and crop insurance programs insure different types of property, with different time horizons. While state insurers and NFIP insure permanent structures that are designed to last years or decades, federal crop insurance primarily covers agricultural commodities that farmers plant and harvest each year. This annual cycle can allow farmers to adapt their insured property to changes in weather-related risk more easily than state or NFIP policyholders can adapt their permanent structures to changing wind or flood risks.

Uncertainty about the magnitude, timing, and extent of the effects of climate change in the future presents challenges to both public and private insurers. For example, NFIP will likely be affected by future sea-level rise. According to the May 2014 National Climate Assessment, although global mean sea level may increase anywhere from 8 inches to 7 feet in this century, the magnitude of the projected future sea-level rise varies considerably along the U.S. coastline due to a variety of factors, such as land subsidence and uplift.[19] Officials from public insurers and some representatives from private industry stated that uncertainty regarding climate projections presents challenges to these insurers' ability to plan for the future.[20] However, as stated in a 2010 National Research Council (NRC) report, even though uncertainties exist regarding the exact nature and magnitude of impacts, mobilizing now to increase the nation's resilience can be an insurance policy against future climate change risks.[21]

We define resilience as the ability to prepare and plan for, absorb, recover from, and more successfully adapt to actual or potential adverse events.[22] Hazard mitigation—actions that reduce the long-term risks of life and property by lessening the impact of disasters—and climate change adaptation—the adjustments to natural or human systems in response to actual or expected climate change—promote resilience to extreme weather events, among other things.

Recent executive orders have addressed the topics of vulnerabilities to extreme weather events and climate change-related risks. Executive Order 13632, which was signed in December 2012, established the Hurricane Sandy Rebuilding Task Force to, among other things, assess current vulnerabilities to extreme weather. The task force was also to identify opportunities for achieving rebuilding success and improving the affected region's resilience, consistent with the National Disaster Recovery Framework's commitment to support economic vitality, enhance public health and safety, protect and enhance natural and man-made infrastructure, and ensure appropriate accountability. Executive Order 13653, which was signed in November 2013, directs federal agencies to, consistent with their missions, (1) address barriers to the nation's resilience to climate change; (2) reform policies that may, perhaps unintentionally, increase the vulnerability of natural or built systems, economic sectors, natural resources, or communities to climate change; and (3) identify opportunities to support and encourage smarter, more climate-resilient investments.[23]

Growth in Federal and Private Sector Exposure Has Increased Losses to Date, and Climate Change May Increase Losses in Coming Decades

Growing federal and private sector exposure since our 2007 report on flood and crop insurance has increased insured and uninsured losses to date, and climate change and related increases in the frequency and severity of extreme weather events may further increase such losses in coming decades.

Federal and Private Sector Exposure Have Grown and Increased Losses to Date

Federal and private sector exposure to potential insured losses grew since 2007. Specifically, inflation-adjusted federal exposure to potential insured losses grew from $1.3 trillion to $1.4 trillion (8 percent) from 2007 through

2013. According to our analysis of the most recent data available, private sector exposure grew from an estimated $60.7 trillion to $66.5 trillion (10 percent) from 2007 through 2012, in 2014 dollars.[24] Property insured under the NFIP comprised 91 percent of federal exposure to insured loss in 2013, but it grew the least (4 percent) from 2007 to 2013. Property insured under the federal crop insurance program accounted for 9 percent of total federal exposure to insured loss in 2013, but it grew the most (68 percent) since 2007, as shown in table 1.

Table 1. Federal Exposure to Potential Loss from Insured Property, 2007 to 2013, in 2014 Dollars

Dollars in billions				
	Total value of property insured		Percentage of federal exposure	Percentage increase
Program	2007	2013	2013	2007-2013
Federal				
National flood insurance	$1,267	$1,321	91%	4%
Federal crop insurance	$75	$125	9%	68%
Total	$1,342	$1,446	100%	8%

Sources: GAO analysis of Federal Emergency Management Agency and Risk Management Agency data on total exposure. | GAO-15-28.

For historical context, in comparing the 7-year period shown in the table to the preceding 7-year period (2000 to 2006), NFIP's exposure to loss grew more slowly from 2007 to 2013, and the federal crop insurance program's exposure grew more quickly.[25] From 2000 to 2006, inflation-adjusted NFIP exposure grew from $749.2 billion to $1.20 trillion (60 percent), compared with 4 percent from 2007 to 2013. In contrast, from 2000 to 2006, the federal crop insurance program's exposure grew from $45.5 billion to $56.9 billion (25 percent) compared with 68 percent from 2007 to 2013, adjusted for inflation. Also, as a share of total federal exposure, in 2006, federal crop insurance comprised 5 percent of total federal exposure, compared to 9 percent in 2013.

Disaster relief appropriations—which could be considered a proxy for federal exposure to uninsured losses—also grew.[26] Based on a 2013 analysis of disaster relief appropriations by the Congressional Research Service, the amount of inflation-adjusted disaster relief per fiscal year increased from a median of $6.2 billion for the years 2000 to 2006, to a median of $9.1 billion for the fiscal years 2007 to 2013 (46 percent).[27]

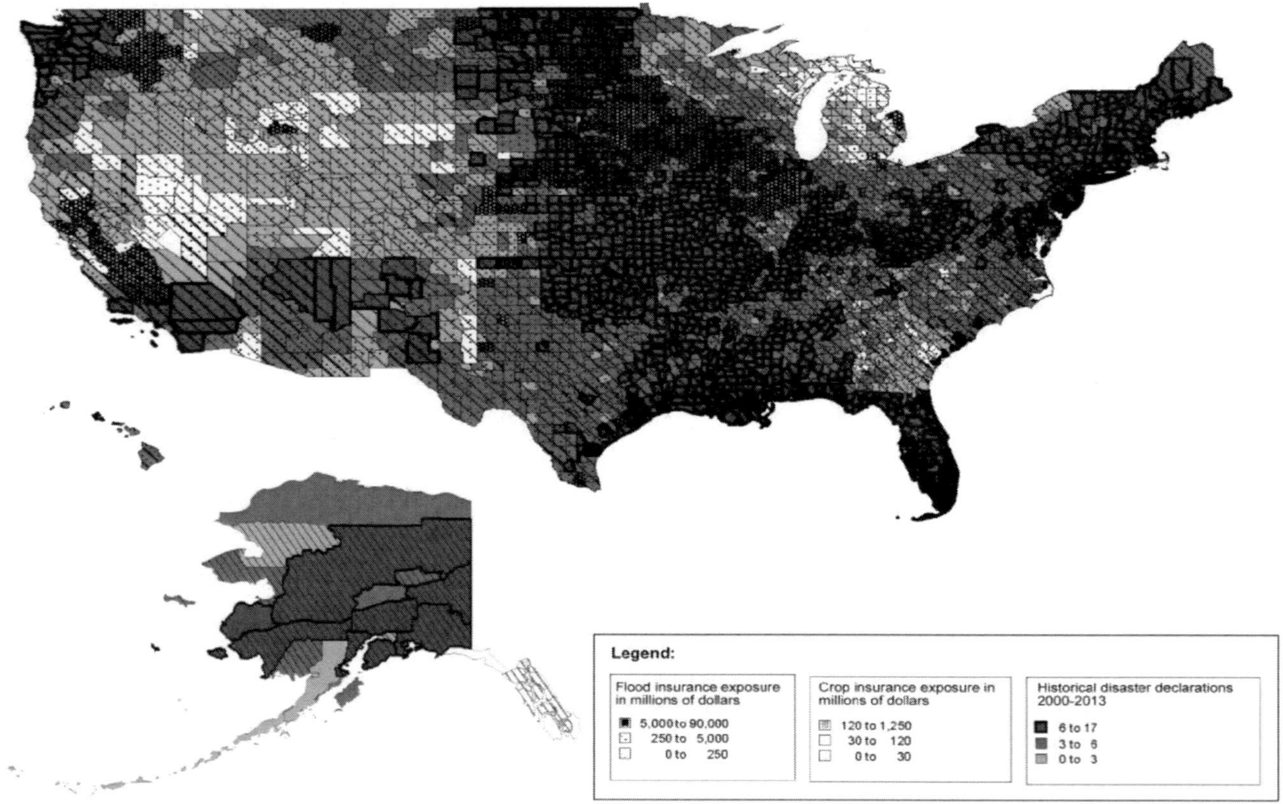

Source: GAO analysis of Federal Emergency Management Agency and Risk Management Agency data; Mapinfo (map). | GAO-15-28.

Notes: We included disaster declarations from 2000 to 2013 to better represent geographic exposure, given the annual variability of weather-related events. 137 out of 15,457 declarations (0.9 percent) during this period were not weather-related and include events such as earthquakes and tsunami-related flooding. The flood insurance program, administered by the Federal Emergency Management Agency, within the U.S. Department of Homeland Security, makes flood insurance available to residential property owners and businesses. According to USDA's website, the federal crop insurance program promotes the economic stability of agriculture through a sound system of crop insurance and providing the means for the research and experience helpful in devising and establishing such insurance. Management is vested in a Board of Directors, subject to the general supervision of the Secretary of Agriculture.

Figure 1. Disaster Declarations 2000-2013, Federal Flood Insurance and Federal Crop Insurance 2013 Exposure by County.

Although federal exposure spans all 50 states, it is concentrated in certain parts of the country, such as the Mississippi River Basin, California's central valley, and coastal areas, as shown in figure 1.

Federal and private insured exposure to loss grew for a variety of reasons, including increases in the value of property insured and increases in the amount of coverage written. For example, according to data that the National Oceanic and Atmospheric Administration (NOAA) obtained from the U.S. Census Bureau, the U.S. coastal population grew by 39 percent from 1970 to 2010, and population density in coastal areas is six times greater than inland areas. Furthermore, a study by an insurance industry modeling firm found that the total value of property in U.S. East Coast and Gulf Coast areas grew by just under 4 percent each year since 2007 to over $10 trillion in 2012, in nominal dollars. Some studies and our prior work, suggest that the current level of federal fiscal exposure to losses may become increasingly difficult to sustain in coming decades, given these socioeconomic factors and other budget constraints.[28]

The 20 scientific and industry studies we reviewed that examined the historical loss record generally found that the exposure growth in hazard-prone areas has largely driven increased insured and uninsured losses to date. Specifically, most loss analyses we reviewed that found an upward trend, identified socioeconomic factors, such as growth in population and the value of property, as the primary drivers of increasing losses to date. Most studies we reviewed did not find a statistically significant increase in such losses conclusively attributable to climate change.[29] One assessment of loss studies noted that climate change cannot be ruled out as a factor because of limitations in data quality, different methods of correcting for socioeconomic trends, and changes in insurance coverage.[30] Although most of the studies we reviewed did not find a clear climate change signal in historical losses, they noted that climate change may start to affect losses in the near future.

Climate Change and Related Extreme Weather May Increase Insured and Uninsured Losses in Coming Decades

Recent assessments of loss projections for certain weather events suggest that climate change may increase losses substantially by 2040, and potentially double annual losses by 2100, compounding existing loss trends. For example, our analysis of 20 scientific studies shows a wide range of projections that, on average, predict a 14 to 47 percent increase in inflation-adjusted U.S.

hurricane-related losses—which significantly contribute to total U.S. losses—attributable to changes in the severity of storms by 2040, and a 54 to 110 percent increase in losses by 2100, as shown in table 2.

Table 2. Estimated Percentage Change in Inflation-Adjusted U.S. Hurricane Insured and Uninsured Losses Potentially Attributable to Climate Change Based on Peer-Reviewed Studies Conducted Since 2007

	Percentage change from 2000 to 2040		**Percentage change by end of century (year)**	
Study author, date	**Range in losses**	**Central estimate in losses**	**Range in losses**	**Central estimate in losses**
Hallegatte, 2007	NA	22%[a]	NA	54% (2100)
Bender et al., 2010	-27 to 36%	14%[a,b]	-54 to 71%	NA[b,c] (2090)
Nordhaus, 2010	12 to 92%	47%[a]	30 to 220%	110% (2100)
Mendelsohn et al., 2012[d]	NA	NA	54 to 240%	93% (2100)

Sources: Adapted from Bouwer, L.M., Projections of Future Extreme Weather Losses Under Changes in Climate and Exposure, Risk Analysis, vol. 33 (2013) 915-930; McAneney, J., et al., Market-Based Mechanisms for Climate Change Adaptation: Assessing the Potential for and Limits to Insurance and Market-Based Mechanisms for Encouraging Climate Change Adaptation, National Climate Change Adaptation Research Facility (Gold Coast, Australia: 2013); Hallegatte, S., The Use of Synthetic Hurricane Tracks in Risk Analysis and Climate Change Damage Assessment. Journal of Applied Meteorology and Climatology, vol. 46, (2007) 1956-1966; and Mendelsohn et. al., "The impact of climate change on global tropical cyclone damage." *Nature Climate Change,* vol. 2, (2012) 205-209. | GAO-15-28

Notes: These estimates are based on different assumptions and methods, including different models of climate change's potential effects. Uncertainty remains about the potential feedback effects in various earth systems that may affect the magnitude of changes in hurricane severity. The end of century estimates extrapolated expected damages based on different time periods: Hallegate and Bender et al. (1900-2005), Nordhaus (1900-2008), and Mendelsohn et al. (1960-2009).

[a]These estimates were scaled to a linear trend from estimates covering different time periods in Bouwer, 2013. We excluded two studies included in Bouwer, 2013, because they were not peer-reviewed.

[b]This estimate measures damage potential rather than expected damage, which requires differentiating between hurricanes that strike the coast and those that remain offshore.

[c]Although an average value of 28 percent is reported in McAneney et al., we could not verify the average because we did not have sufficient information from the source to independently verify the average.

[d]Results are for North America, not the U.S. alone.

With exposure projected to increase over the same period, annual insured and uninsured losses could be much higher by 2100 based on some of the studies we reviewed.[31] Specifically, one peer-reviewed study projected property exposure in hazard-prone areas to more than double losses by 2100.[32] When combined with the range of climate change projections, total losses could increase anywhere from about 50 to 340 percent by 2100.

For agriculture, climate disruptions to production have increased in the past 40 years, and the May 2014 National Climate Assessment projects such disruptions will increase over the next 25 years. According to this report, producers have many available strategies for adapting to the average temperature and precipitation changes projected for the next 25 years, including technological advancements, expansion of irrigated acreage, and regional shifts in crop production, among others. However, according to the report, by midcentury, when temperatures could increase by between 1.8°F and 5.4°F and precipitation extremes could further intensify, expected yields of major U.S. crops and farm profits could decline, even with the current pace of technological advances and geographic shifts.[33]

Public Insurers Have Begun Taking Steps to Prepare for Climate Change Effects, but Challenges May Limit Their Ability to Manage Future Exposure

Public insurers have commissioned climate change studies, incorporated climate change adaptation into their planning, and taken other steps to better understand and prepare for climate change's potential effects. However, inherent challenges of federal insurance programs, such as how federal insurers can address policyholders' long-term risk given the short-term focus of insurance contracts, may impede NFIP and RMA's ability to minimize long-term federal exposure to climate change.

Public Insurers Have Begun Taking Steps to Understand Climate Change Risk and Prepare for Related Effects

Public insurers have begun taking steps to understand climate change risks, such as sea-level rise, and have identified actions that could manage their exposure to climate change's effects. For example, federal insurers commissioned climate change studies and modeling to better understand the long-term implications of climate change for their programs. Regarding flood insurance, a FEMA-commissioned June 2013 study found that sea-level rise, erosion, and other changes in coming decades will affect the NFIP by expanding areas prone to flooding and requiring premium increases to cover higher losses.[34] The report also found that small annual rate increases could allow NFIP to adjust to gradual climate change effects. Based on this report's findings, FEMA initiated two pilot studies to analyze sea-level rise and its impacts in special flood hazard areas. FEMA appointed members to a Technical Mapping Advisory Council to, among other things, develop recommendations to ensure FEMA uses the best available methodology to consider future development on flood risk as required by the Biggert-Waters Flood Insurance Reform Act of 2012 (Biggert-Waters Act).[35] According to FEMA officials, the council expects to release its flood mapping report in September 2015.

Regarding crop insurance, RMA's 2009 study found that the crop insurance program has self-correcting features, which allow RMA to manage its exposure to gradual changes in climate.[36] For example, farmers' annual production history determines their coverage and, if yields decrease under climate change, the program's exposure adjusts downward as well. RMA may also adjust the availability of coverage for crops to respond to geographic shifts in production as conditions become more or less favorable for certain crops. According to RMA officials, farmers' desire to maximize their profits and maintain their businesses will motivate them to alter their production practices as climate change effects occur. However, the agency's 2009 study also recommended that RMA should better monitor weather and climate to understand potential rapid changes in future conditions and how to adapt agricultural and risk management practices to address climate change. RMA entered into a cooperative agreement with Oregon State University's Parameter-elevation Regressions on Independent Slopes Model (PRISM), Climate and Weather Group that will enhance RMA's monitoring of weather and climate.[37] Representatives from PRISM and RMA said the two organizations have engaged in a long-term program to develop a 100-year historical weather time series that could help analyze climate change's effect on agriculture.

Furthermore, in response to recent executive orders, FEMA and RMA have developed climate change adaptation plans, which outline planned agency actions to manage significant climate change related risks to, and vulnerabilities in, agency operations and missions.[38] For example, according to its climate change adaptation policy statement, FEMA will continue its study of climate change impacts on NFIP. Furthermore, according to program officials, climate change efforts will incorporate the best climate science available into flood maps—which form the basis for identifying property owners' flood risk and providing guidance to communities on land-use decisions. According to its adaptation plan, RMA will update program parameters, such as the earliest and final planting dates. Changing planting dates, as needed, can help farmers avoid exposing crops to new changes in weather or climate. RMA officials told us that they also recently worked with USDA's Natural Resources Conservation Service to ensure consistency between crop insurance and conservation programs that protect soil and water resources.

RMA also participates in the broader USDA climate change effort, such as USDA's Global Change Task Force, according to RMA and USDA officials. This task force coordinates USDA activities related to climate change and provides a venue for sharing information within the agency, according to an agency document. In addition, USDA's Agricultural Research Service (ARS), has identified several climate change adaptation strategies to promote long-term resilience to climate change effects such as increased soil erosion and water scarcity. Specifically, a 2013 ARS report found that resilient agricultural practices such as conservation tillage—where farmers leave some crop residue on fields— and water conservation will help minimize climate change costs and sustain agricultural production in a changing climate. According to USDA officials, the agency also plans to develop region-specific adaptation strategies through its newly established Regional Climate Hubs, which are to deliver science-based, practical information to farmers and to support decision making related to mitigation of, and adaptation to, climate change.

Additionally, for a variety of reasons, federal insurers have adjusted their rate-setting calculations in ways that may better position them to respond to climate change. Although not specific to climate change, the Biggert-Waters Act requires NFIP to establish a reserve to help meet expected future obligations and to establish standards that ensure that its flood maps' flood risk determinations are adequate, both of which should help the agency collect funds that more closely match the risks it incurs or that are more likely to be sufficient to cover losses. FEMA officials said the agency has also increased the reserve amount in premiums to cover flood risk uncertainty, to better reflect the current flood risk associated with properties below base flood elevation. Recently enacted legislation required FEMA to phase out subsidies for some properties such as nonprimary residences and business structures through 25 percent annual premium increases until the full-risk rate is reached.

In response to a 2011 rate study, RMA changed its rate calculations to more heavily weigh recent weather data, which RMA documents suggest could enable premiums to more quickly and fully reflect changes in the climate.[39] By law, RMA must set crop insurance premiums at rates sufficient to cover projected claims. Due in part to weather variability, in some years, RMA will likely collect more premiums than it pays in claims, and vice versa. RMA currently uses an average of the previous 20 years of yields to calculate its rates, rather than the previous method of assigning equal weight to all years back to the fixed base year of 1975— which was in place prior to 2012.[40]

Moreover, FEMA, in conjunction with other federal agencies, has taken some recent steps to manage future risks related to climate change for disaster relief. For example, the Hurricane Sandy Rebuilding Task Force implemented a minimum flood risk reduction standard for Sandy-related disaster funding to account for future sea-level rise in response to Executive Order 13632.[41] Under this standard, structures repaired or rebuilt must meet forward-looking standards, such as elevating the ground floor of a building 1 foot higher than existing FEMA standards. In addition, according to FEMA officials, a current interagency effort seeks to develop a Federal Flood Risk Reduction Standard that would apply to future disaster relief appropriations—although it is too early to know whether such a standard will incorporate future risk. Furthermore, according to a July 2014 White House statement, FEMA will issue new guidance that requires states to incorporate climate change into their hazard mitigation plans as a condition for receiving disaster relief.[42] FEMA's Hazard Mitigation Assistance programs and post-disaster grants currently do not require grantees to incorporate sea-level rise into their cost-benefit calculations for proposed projects, although they do allow it.[43]

Regarding other public insurance programs, at the state level, some state insurers in hazard-prone areas have transferred risk to the private sector to reduce their exposure to claims. Specifically, officials at two of the state insurers we interviewed told us they have sold policies ("depopulation"), bought reinsurance to transfer a portion of the risk in their portfolio to another insurer, and sold bonds that provide funding to the insurer should a catastrophic weather event occur ("catastrophe bonds"). These risk transfer methods reduce exposure to losses from extreme weather. Reinsurance and catastrophe bonds are short-term, market-based risk transfer methods, typically of 1 to 3 years in duration.[44] We previously concluded that there are several strategies—many which would require statutory authority to implement but a few that FEMA currently has authority to implement—to be considered at the federal level to allow the transfer of risk to the private sector.[45] The Biggert-Waters Act also required FEMA to issue a report in 2013 that assessed the capacity of the private reinsurance, capital, and financial markets to assume a portion of the insurance risk of NFIP.

Federal Insurers Face Challenges Managing Future Exposure to Climate Change

Federal insurers face two main program challenges that may constrain their ability to manage their fiscal exposure and address future climate change risk. First, federal law encourages federal insurers, such as FEMA and RMA, to provide affordable insurance to policyholders through subsidized rates, which lessens the agencies' ability to collect sufficient premiums from policyholders to pay claims, increases the federal government's fiscal exposure, and may reduce policyholders' incentives to manage risk by giving them inaccurate signals about the level of risk.[46] Specifically, federal insurers face a challenge between providing affordable premiums through subsidies and managing financially self-sufficient programs by charging policyholders full-risk premiums. Additionally, while insurers, in general, communicate the risk of incurring losses to policyholders through their premium rates, by subsidizing some policies, federal insurers have not provided all policyholders with accurate price signals about their risk of incurring losses.[47] As a result, some NFIP and federal crop insurance policyholders may perceive their risk of loss to be lower than it really is and may have less financial incentive to reduce this risk.[48]

For example, FEMA offers subsidized premium rates for policies covering certain structures, some of which are in high-risk areas. As a result of NFIP's historical rate structure, the program has generated sufficient premiums to cover claims in years with average losses but has not had sufficient funds to cover claims in catastrophic loss years, and FEMA has an outstanding balance with the U.S. Treasury of $24 billion to pay for claims in these years. Although FEMA is phasing out most subsidies, they may remain in place for many years, and while they are in place, newly purchased residential properties may qualify for subsidized NFIP rates.[49] As long as some NFIP policyholders only pay for a portion of their risk of losses due to subsidized premiums, they receive inaccurate price signals about their property's full flood risk—regardless of what other information FEMA may provide. Consequently, policyholders who receive subsidized rates may have limited financial incentive to take steps, such as floodproofing their homes above base flood elevation, to reduce their risk.[50]

For federal crop insurance, although RMA is required to collect sufficient premiums to cover projected claims, the premiums are subsidized by the federal government. For example, we previously found that the government paid about 62 percent of premium costs for the program in 2011. The government continued to pay 62 percent of premium costs in 2013, which totaled about $7.3 billion.[51] Also, in August 2014, we found that the costs of federal crop insurance have grown, primarily due to an increase in the value of premium subsidies.[52] According to an April 2014 Congressional Budget Office report, crop insurance program costs are expected to average $8.9 billion annually, for fiscal years 2014 through 2023. Similar to flood insurance, federal crop insurance policyholders receive inaccurate price signals about their potential risk of loss when they receive such premium subsidies. Although farmers are informed of the subsidy amount and have an incentive to maximize their annual yields, they do not bear the true cost of their risk of loss due to weather-related events, such as drought—which could affect their farming decisions.[53]

In prior work, we concluded that reducing subsidies and charging full-risk premiums to individual policyholders would decrease the federal government's fiscal exposure under the flood and crop insurance programs.[54] Reducing the federal government's fiscal exposure to losses under federal insurance programs and sending more accurate signals to policyholders about their risk, become even more important as the risks from climate change and related extreme weather events increase.

Second, given the short-term nature of insurance, public insurers face a challenge in encouraging their policyholders to reduce their long-term exposure to climate change risks.[55] Property insurance contracts typically estimate and communicate risk of property losses for the 1-year term of a policy. However, climate change effects on insured property, such as buildings or fields in production, may span decades. Because FEMA and RMA only provide policyholders with price signals for expected losses in the upcoming year, these policyholders may not be encouraged to reduce long-term risks for their property. Similarly, in agriculture, the tension between the short- and long-term extends beyond insurance into farming operations. According to RMA officials, farmers make annual decisions on the starting times and types of crops they plant and buy seed for the following year. They focus more heavily on this short-term time horizon, in part because seed and farming technology constantly change.

Furthermore, building to community standards that are identical to existing NFIP standards—which are based on near-term flood risk—may unintentionally increase policyholders' long-term vulnerability to climate change as sea-level rises or erosion increases properties' flood risk— which does not reflect the direction contained in Executive Order 13653.

This is because, in the short-term, policyholders may be less likely to consider taking actions to protect their homes or localities against future risk and may continue to build in high-risk areas. According to FEMA guidance, the NFIP standards are periodically revised to incorporate new regulations or clarify old ones. Communities in turn must then update their own regulations to maintain consistency with the NFIP standards. Without incorporating forward-looking minimum standards into NFIP's requirements, similar to the minimum standard applied by the Hurricane Sandy Rebuilding Task Force, NFIP policyholders and localities may continue to build and rebuild structures to current community standards that may not reflect the changing weather-related risks faced over structures' designed life spans—thereby exacerbating the federal financial risk to climate change.

Regarding the federal crop insurance program, federal law prohibits crop insurance from covering losses due to farmers' failure to follow good farming practices. RMA's good farming practices provide acceptable farming methods for crop insurance policyholders to use in producing yields consistent with historical production. However, these practices are focused on maintaining historic crop yields over the term of the annual insurance contract, and some of these practices may unintentionally increase the vulnerability of agriculture to climate change, contrary to Executive Order 13653's directive for agencies to manage vulnerabilities to climate change. For example, certain practices, such as conventional tillage and traditional irrigation methods, may maintain historic crop yields in the short-term, but they may inadvertently reduce agriculture's long-term resilience through increased erosion, depleted soil quality, and inefficient water use. RMA officials said that crop insurance adjusts to farmers' best management practices and should follow their adjustments rather than dictate their behavior. However, a variety of factors influence farmers' production practices and, in some cases, farmers may not adopt resilient practices unless these practices also maximize farmers' short-term net benefits, consistent with crop insurance's focus on maintaining historic production.[56] By not encouraging agricultural experts to recommend or incorporate resilient agricultural practices into their expert guidance for growers' good farming practices, RMA is likely missing an opportunity to decrease existing and future fiscal exposures to climate change. Consequently, crop insurance may continue to cover losses resulting from practices that increase vulnerability to climate change.

MANY PRIVATE INSURERS AND REINSURERS HAVE IDENTIFIED AND ASSESSED CLIMATE CHANGE RISKS, BUT PREPARING FOR LONG-TERM EFFECTS IS CHALLENGING

Many private property and casualty insurers and reinsurers have taken some steps since our 2007 report on flood and crop insurance to understand and report on their risks associated with climate change, including participating in industry climate change surveys, and issuing reports that identify and assess climate change risks and trends in weather-related losses. While selected insurers we interviewed said they manage climate change risks through their underwriting practices, we found that the industry faces challenges in preparing for long-term climate change, such as short-term insurance contracts and catastrophe modeling limitations.

Many Private Insurers and Reinsurers Identified and Assessed the Climate Change Risks They May Face

Since our 2007 report on flood and crop insurance, industry representatives, including insurers, reinsurers, and brokerage firms, have reported on climate change risks they face. For example, many insurers have reported on how they incorporate climate change into their risk management practices through an annual industry survey adopted by the National Association of Insurance Commissioners (NAIC) in 2010.[57] According to an analysis by the California Department of Insurance, which administers the survey, 1,067 insurers participated in the climate risk disclosure survey in 2013, for reporting year 2012.[58] The California Department of Insurance found that 74 percent of insurers that participated in the 2012 survey indicated that they have a process for identifying and assessing climate change-related risks, and 65 percent of insurers indicated that they have encouraged policyholders to reduce the losses caused by climate change-influenced events. According to several insurers and industry representatives we interviewed, insurers generally manage their exposure to weather-related losses associated with climate change through their underwriting practices, which include charging risk-based premiums, managing coverage options, and sharing exposure to losses by purchasing reinsurance.

In addition to insurers' participation in the climate risk disclosure survey, other insurance industry representatives have identified and assessed risks associated with climate change. For example, representatives from three of four reinsurers we interviewed said that they have issued reports on climate change risks or weather-related losses. In addition, one of two reinsurance brokers we interviewed has reported on climate change risks and scientific assessments. The other reinsurance broker we interviewed has issued monthly and annual reports on natural catastrophe losses worldwide based on publicly available industry and climate data that cover topics associated with climate change, including trends in hurricane frequency, global temperatures, and weather-related losses.

Preparing for Long-Term Climate Change Effects is Challenging for Private Insurers

Private insurers face several challenges to prepare for the long-term effects of climate change. One key challenge is similar to that identified by public insurers: the short-term nature of insurance contracts. Insurers generally write property coverage for 1-year terms. While short-term contracts allow insurers flexibility to manage their exposure to losses through (1) changes in coverage or pricing or (2) not renewing a particular policy, the annual period of each contract makes it challenging for insurers to incorporate long-term climate change projections into their risk management practices.

According to several industry representatives we interviewed, insurers and reinsurers use advanced computer modeling, known as catastrophe models, to help estimate risks and price insurance policies. While catastrophe modeling helps insurers and reinsurers estimate potential short-term weather-related losses, they said that incorporating the effects of climate change into catastrophe models remains a challenge. According to a representative of a catastrophe modeling firm we interviewed, catastrophe models that insurers use to price policies generally estimate short-term risks based on historical weather data and past losses and not on long-term climate change projections.[59] Many industry representatives said that advances in catastrophe modeling have improved risk measurement over time. However, estimating weather-related risks still includes elements of uncertainty, and catastrophe modeling information is limited for some weather-related risks. We interviewed representatives from two catastrophe modeling firms and a reinsurance broker that have developed models to estimate risks associated with hurricanes, including wind and storm surge impacts, as well as a model for estimating the effects of weather on crops. Several firms are developing catastrophe models for inland flood risk, which will help estimate potential flood risks, although FEMA officials and several industry representatives said these models are not yet advanced enough to allow insurers to estimate or price flood risk for individual properties.

Even with these challenges, several industry representatives said that, barring any additional regulatory restrictions, insurers and reinsurers are positioned to continue managing risks associated with climate change through their ability to set risk-based prices, write coverage, or manage exposure to losses. However, two reports by an industry group and academics have found that past catastrophic events have caused the private property casualty insurance market to retract following a major disaster, placing greater pressure on public insurers to provide coverage. For example, according to a 2012 Insurance Information Institute report, some insurers became insolvent or stopped writing coverage in certain areas following Hurricane Andrew in 1992 and, since then, Florida's state insurer has grown from a market of last resort to the state's largest insurer.[60] While some industry representatives we interviewed said insurers and reinsurers have successfully managed weather-related risks through underwriting, some said additional incentives are needed to help the private sector, government programs, and individuals manage their exposure to risks associated with climate change.[61] For example, several industry representatives have said that South Carolina's tax credit program for homeowners who fortify their homes to make them more resistant to hurricanes, catastrophic wind events, or flooding could encourage individuals to take steps to reduce their exposure to such risks.[62]

Some industry representatives we interviewed also suggested that federal, state, and local government adoption of building and land use practices that recognize potential climate change effects would help decrease both public and private exposure to insured and uninsured weather-related losses.[63] For example, land use practices and zoning regulations that recognize potential climate change impacts could help reduce public and private exposure to climate change by limiting new construction or relocating existing structures from hazard-prone areas. In addition, several insurers and reinsurers we interviewed and other industry representatives have suggested that incentives are needed to encourage more state or local governments to adopt resilient building standards to help mitigate against weather-related losses. The Hurricane Sandy Rebuilding Task Force reported in August 2013 that using disaster-resistant building codes is the most effective method to ensure new and rebuilt structures are designed and constructed to a more resilient standard.[64] The task force recommended that states and local governments adopt the most current building codes to ensure that buildings and other structures incorporate the latest science, advances in technology, and lessons learned. A representative from one industry group said updated and more resilient building codes, as well as improved enforcement measures, would help reduce exposure to weather-related risks including hurricanes, floods, wildfires, hail, and wind storms, which are associated with climate change. The industry group reported in December 2011 that while several hurricane-prone states, including Florida, Virginia, and New Jersey, have adopted more resilient building codes, opportunities exist for others to adopt stronger standards and better enforcement measures.[65]

To examine these and other issues, in November 2013, the President established a Task Force on Climate Preparedness and Resilience composed of state, local, and tribal leaders to advise the President and an interagency council on how the federal government can support state, local, and tribal preparedness for and resilience to climate change, among other things. The task force's recommendations, expected in the fall of 2014 according to a White House press release, will address removing barriers to resilient investments—such as improving data and tools available to state and local decision makers, reforming existing policies and federal funding programs, and identifying opportunities to support more climate-resilient investments by states, local communities, and tribes.[66]

CONCLUSION

FEMA and RMA have commissioned climate change studies, incorporated climate change adaptation into their planning, and taken other steps to address aspects of their federal flood and crop insurance programs that create fiscal exposures associated with climate change and extreme weather.

However, the agencies continue to face fundamental challenges that send inaccurate price signals to policyholders about their potential risk of loss and increase federal fiscal exposure, and may unintentionally increase their policyholders' vulnerability to climate change risks. We have previously concluded, among other things, that reducing subsidies and charging full-risk premiums to individual policyholders would decrease the federal government's fiscal exposure under the flood and crop insurance programs.

Executive Order 13653 directs federal agencies to, consistent with their missions, reform policies that may, perhaps unintentionally, increase the vulnerability of natural or built systems, economic sectors, natural resources or communities to climate change. Regarding FEMA's flood insurance program, the agency is phasing out most subsidies, and is studying how to incorporate the projected effects of climate change, such as future sea-level rise and erosion, into its flood maps, but the mapping advisory council's recommendations aren't expected until September 2015. Until the agency implements these changes, some NFIP policyholders will continue to receive inaccurate signals about their current risk of loss, and all may not receive signals about their future risk of loss over the designed lifespan of their insured property. Furthermore, NFIP standards may not fully reflect policyholders' long-term vulnerability to climate change because these standards are based on current risk that does not reflect future sea-level rise, erosion, or other changes. Without incorporating forward-looking minimum standards into NFIP's construction and rebuilding requirements, similar to the minimum standard adopted by the Hurricane Sandy Rebuilding Task Force, NFIP policyholders and communities may continue to build and rebuild structures to current NFIP standards that do not necessarily reflect the changing weather-related risks faced over structures' designed life spans—which could exacerbate federal fiscal exposure amid already strained federal resources.

In addition, a variety of agricultural practices are available to farmers that would improve their long-term resilience to climate change, such as practices that would promote long-term water conservation and soil conservation. However, federal crop insurance policyholders may receive inaccurate price signals about their current risks because of subsidized premiums and because of the short-term nature of annual insurance contracts they may not receive signals that reflect the long-term implications of their short-term farming practice decisions. Additionally, federal law prohibits crop insurance from covering losses due to a farmers' failure to follow good farming practices, although some of these practices may increase the vulnerability of agriculture to climate change, which may not reflect the direction contained in Executive Order 13653. Without working with agricultural experts to recommend or incorporate resilient agricultural practices into their expert guidance for growers' good farming practices, RMA is likely missing an opportunity to decrease existing and future fiscal exposures to climate change. Consequently, crop insurance may continue to cover losses resulting from practices that increase vulnerability to climate change.

RECOMMENDATIONS FOR EXECUTIVE ACTION

We are making two recommendations in this report.

- To promote forward-looking construction and rebuilding efforts while FEMA phases out most subsidies, we recommend that the Secretary of the Department of Homeland Security direct FEMA to consider amending NFIP minimum standards for floodplain management to incorporate, as appropriate, forward-looking standards, similar to the minimum standard adopted by the Hurricane Sandy Rebuilding Task Force.
- To promote greater resilience to climate change effects in U.S. agriculture, we recommend that the Secretary of Agriculture direct RMA to consider working with agricultural experts to recommend or incorporate resilient agricultural practices into their expert guidance for growers, so that good farming practices take into account long-term agricultural resilience to climate change.

AGENCY COMMENTS AND OUR EVALUATION

We provided a draft of this report to USDA, Commerce, DHS, and Treasury for review and comment. USDA provided written comments; Commerce provided technical comments, which we incorporated as appropriate; DHS provided written comments, and the Department of the Treasury did not provide comments. USDA did not specify their agreement or disagreement with our recommendation, and DHS agreed with our recommendation.

In its written comments, USDA referenced our finding that RMA's good farming practices focus on maintaining historic crop yields over the term of the annual insurance contract and that some of these practices may unintentionally increase the vulnerability of agriculture to climate change, contrary to Executive Order 13653's directive for agencies to manage vulnerabilities to climate change. USDA stated that RMA does not direct producers to take or carry out certain agronomic practices, but rather relies on guidance from agricultural experts in the area. USDA also stated that to the extent that agricultural experts in an area recommend or incorporate resilient practice recommendations into their expert guidance for growers in a given area, then RMA would consider such in its good farming practice determination for coverage of losses. While RMA may not direct producers to follow certain agronomic practices, it can provide incentives for farmers' adoption of resilient practices by working with agricultural experts to recommend or incorporate resilient practices into their expert guidance for growers' good farming practices and therefore eligibility determinations for claim payments. As we note in the report, RMA has an opportunity to potentially reduce agriculture's long-term vulnerability to climate change by encouraging the adoption of resilient practices now. For that reason, we recommend that RMA consider working with agricultural experts to recommend or incorporate resilient agricultural practices into their expert guidance for growers so that good farming practices take into account long-term agricultural resilience to climate change.

In addition, USDA refers to one of our identified program challenges to insurers that federal law encourages them to provide affordable insurance to policyholders through subsidized rates, which may reduce policyholders' incentives to manage risk by giving them inaccurate signals about the level of risk. USDA states that, while the federal government does subsidize a significant share of a producer's premium, every producer receives a notification from his or her insurance provider, which explains how much premium was paid by the government and how much is to be paid by the producer. Therefore, USDA states that all producers are made aware of the full risk. While notifying farmers of the subsidy amount provides useful information, farmers do not bear the true cost of their risk of loss. As a result, the market signal sent by federal insurers for the price of the policyholders' risk is the amount the policyholders actually pay. We have modified the report and our recommendation to respond to USDA's comments.

In its written comments, DHS concurred with our findings that FEMA has taken action to better understand and prepare for climate change's potential effects and that FEMA faces challenges that may limit its ability to minimize long-term federal exposure to climate change.

DHS also concurred with our recommendation to consider amending NFIP minimum standards for floodplain management to incorporate, as appropriate, forward-looking standards, similar to the minimum standard adopted by the Hurricane Sandy Rebuilding Task Force. In particular, DHS stated that FEMA has already taken action to consider amending NFIP minimum standards by (1) commissioning a 2006 study to assess the adequacy of the NFIP's minimum standards, (2) participating in a 2004 forum with key stakeholders about amending the NFIP minimum standards, and (3) encouraging communities to participate in the NFIP's Community Rating System, which offers discounted flood insurance rates in exchange for a community's proactive efforts to reduce its flood risk.[67]

We do not believe that these actions meet the intent of our recommendation for several reasons. First, commissioning a study and participating in a policy forum that discussed several aspects of NFIP minimum standards are not sufficient evidence that FEMA officials considered adopting forward-looking standards. FEMA has not provided documentation of internal policy discussions or other actions taken in response to the study or forum's findings.

Additionally, the study commissioned by FEMA is 8 years old, and it therefore does not reflect scientists' current understanding of sea-level rise and other climate change effects identified in more recent National Climate Assessments.

Similarly, the discussion with stakeholders occurred 10 years ago, and it therefore does not reflect the current state of the program, stakeholders' current understanding of climate-related risks, the adequacy of FEMA's floodplain maps, and recent advances in mapping technology. Moreover, the Community Rating System is voluntary and, as of March 2014, 1,296 of the nearly 22,000 NFIP-participating communities are in the program. These communities represent about 67 percent of NFIP policyholders. Accordingly, amending the NFIP minimum standards could reach the over 20,000 nonparticipating communities.

As the summary report from the 2004 forum notes, building in a margin of error such as adding a foot or more to the calculated base flood evaluation for flood hazard assessment at the outset of the program could have avoided many of the program's current problems regarding uncertainty—including the uncertainty of climate change.

We continue to believe that FEMA should consider amending the NFIP minimum standards to incorporate, as appropriate, forward-looking standards, and we therefore do not consider the recommendation resolved and closed.

J. Alfredo Gómez
Director, Natural Resources and Environment

APPENDIX I: OBJECTIVES, SCOPE, AND METHODOLOGY

To determine how federal and private sector exposure to losses has changed since our 2007 report on flood and crop insurance and how climate change has or may affect insured and uninsured losses, we analyzed federal and private sector exposure data from 2007 through 2013. To understand longer-term trends in exposure, we also analyzed federal exposure data from 2000, and we reviewed scientific studies and other available literature on climate change. Specifically, we collected agency data on the total value of property insured under the National Flood Insurance Program (NFIP) and the federal crop insurance program. For the private sector, we collected estimates of the total value insured by property-casualty insurers—excluding auto—as determined by two industry catastrophe modeling firms, AIR Worldwide and Risk Management Solutions. We assessed the reliability of the agency and industry data we collected and determined it was sufficiently reliable for describing the total value of property insured.[1] We also conducted a literature review to identify pertinent studies on how climate change has or may affect insured and uninsured losses. Specifically, we searched for scholarly articles, industry articles and reports, think-tank reports, conference reports, and government publications published from 2007 onward. Through the literature search, we identified a number of studies that discussed climate change's potential effects on insured and uninsured losses and stakeholder perspectives of climate change risks. For summarizing the effect of climate change on insured and uninsured losses, we limited our review to scientific, empirical studies that evaluated the historical loss record or projected future losses. Each of these studies used various techniques to discern the respective influences on the loss record of changes in exposure (e.g., wealth, population, insurance penetration) versus changes in weather patterns. Based on these criteria, we identified a total of 64 studies that were relevant and applicable to our report, 20 of which directly addressed the issue of climate change's effect on losses. We reviewed the methodologies of these studies to ensure that they were sound and determined that they were sufficiently reliable for describing the potential effect of climate change on insured and uninsured losses.

To determine how public insurers are preparing for climate change, we reviewed agency documents related to climate change. We reviewed the Department of Homeland Security's (DHS) 2012 Climate Change Adaptation Roadmap, the 2011 Federal Emergency Management Agency's (FEMA) Climate Adaptation Policy Statement, and a 2013 AECOM climate change study which FEMA commissioned. We also reviewed the Risk Management Agency's (RMA) 2012 Climate Adaptation Plan and 2011-2015 Strategic Plan, RMA's cooperative agreement with Oregon State's PRISM group, documents related to the agency's 2011 rate change calculations, and a 2009 RTI climate change and modeling study which the agency commissioned. In addition, at the federal level, we interviewed officials from NFIP, RMA, the U.S. Department of Agriculture, the National Oceanic and Atmospheric Administration, the Federal Insurance Office, and Council on Environmental Quality. We also reviewed the Biggert-Waters Flood Insurance Reform Act of 2012, the Homeowner Flood Insurance Affordability Act of 2014, and the Agricultural Act of 2014. At the state level, we interviewed officials from the Department of Insurance in three states and the wind insurers in two of those three states. We selected a nonprobability sample of states based on the following factors: identified by an industry forecaster as most at-risk to natural disaster, whether the state had an insurance pool administered by a governmental entity or entity established pursuant to state law, and whether the state had experienced an extreme weather-related event in the past decade. Because we used a nonprobability sample to select states, our results are not generalizable to all 50 states; however, our results can provide illustrative information. We also spoke with several academic experts on agriculture and modeling identified from publications, other experts, and our prior work.

Further, to determine how private insurers and reinsurers are preparing for climate change, we reviewed over 20 industry reports and other information from industry representatives, and we interviewed a nonprobability sample of representatives from four insurers and four reinsurers, as well as catastrophe modeling firms, reinsurance brokers, and industry groups representing more than 1,000 large and small property casualty insurers. We selected a sample of four insurers to interview based on market share, experience with flood or crop insurance, and involvement in climate change issues. For insurers selected based on market share, we identified those with a large share of the U.S. property casualty market (about 50 percent cumulative share of the U.S. property casualty market) and others with a smaller share of the market (1 percent or less), based on 2012 industry data. Among these firms, we selected four insurers with diverse experience in the public and private insurance markets, and the four firms we interviewed represented over a 25 percent share of the U.S. property and casualty insurance market, and a 15 percent share of the private crop insurance market. We also interviewed representatives from a sample of 4 reinsurers, from among 10 reinsurance firms with the greatest share of the U.S. reinsurance market. The reinsurers we interviewed represented over a 30 percent share of the U.S. reinsurance market, based on 2012 industry data. In addition, we interviewed 11 other industry participants, as well as academic researchers for context. We identified industry participants to interview through our prior work and relevant publications, as well as suggestions from other interviewees. Industry participants we interviewed included representatives from two catastrophe modeling firms, two reinsurance brokerage firms, and seven industry groups representing insurers, reinsurers, and others. Among the industry groups we interviewed, two groups represented more than 1,000 large and small property casualty insurers. Because we used a nonprobability sample to select interviewees, our results are not generalizable to industry as a whole but provide illustrative examples.

We conducted this performance audit from November 2013 to October 2014 in accordance with generally accepted government auditing standards. Those standards require that we plan and perform the audit to obtain sufficient, appropriate evidence to provide a reasonable basis for our findings and conclusions based on our audit objectives. We believe that the evidence obtained provides a reasonable basis for our findings and conclusions based on our audit objectives.

End Notes

[1] Melillo, Jerry M., Terese (T.C.) Richmond, and Gary W. Yohe, eds., *Climate Change Impacts in the United States: The Third National Climate Assessment*. U.S. Global Change Research Program (Washington, D.C.: U.S. Government Printing Office, 2014). Thomas R. Karl, Jerry M. Melillo, and Thomas C. Peterson, eds., *Global Climate Change Impacts in the United States* (Cambridge University Press: 2009). The 2014 report summarizes the impacts of climate change on the United States, now and in the future. A team of several hundred experts guided by a 60-member Federal Advisory Committee produced the report, which was extensively reviewed by the public and experts, including federal agencies and a panel of the National Academy of Sciences.

[2] GAO, *High-Risk Series: An Update,* GAO-13-283 (Washington, D.C.: February 2013).

[3] FEMA's Disaster Relief Fund (DRF) is the major source of federal disaster recovery assistance for state and local governments when a disaster is declared. FEMA obligates funds from the DRF to help jurisdictions respond to and recover from declared disasters. The DRF is generally funded at a level that is sufficient for what are known as "normal" disasters, with outlays less than $500 million. When a large disaster occurs, additional funding for the DRF may be provided through emergency supplemental appropriations. For more information, see Congressional Research Service, *Disaster Relief Funding and Supplemental Appropriations for Disaster Relief*, R40708 (Washington, D.C.: Aug. 5, 2013) and GAO, *Federal Disaster Assistance: Improved Criteria Needed to Assess a Jurisdiction's Capability to Respond and Recover on Its Own*, GAO-12-838 (Washington, D.C.: Sept. 12, 2012).

[4] GAO-13-283.

[5] FEMA will need to implement provisions in both the Biggert-Waters Flood Insurance Reform Act of 2012, Pub. L. No. 112-141, tit.ll, subtit. A, 126 Stat. 405 (2012) and the Homeowner Flood Insurance Affordability Act of 2014, Pub. L. No. 113-89, 128 Stat. 1020 (2014), that affect many aspects of NFIP, including its finances, rate setting, mapping, and participation, among other things. In addition, over the past 11 years, we have made numerous recommendations to FEMA to improve its administration of the program. However, as we recently reported, FEMA has not fully addressed all of our prior recommendations. See GAO, *Overview of GAO's Past Work on the National Flood Insurance Program*, GAO-14-279R (Washington, D.C.: Apr. 9, 2014).

[6] GAO, *Fiscal Exposures: Improving Cost Recognition in the Federal Budget*, GAO-14-28 (Washington, D.C.: Oct. 29, 2013); GAO-13-283; and GAO-12-838.

[7] GAO, *Extreme Weather Events: Limiting Federal Fiscal Exposure and Increasing the Nation's Resilience*, GAO-14-364T (Washington, D.C.: Feb. 12, 2014); GAO-14-28; and *Budget Issues: Budgeting for Federal Insurance Programs*, GAO/T-AIMD-98-147 (Washington, D.C.: Apr. 23, 1998), According to FEMA officials, FEMA recently increased the requested amount of disaster relief funding during the budget cycle, which may reduce the federal government's fiscal exposure.

[8] GAO-14-364T; GAO-14-28; GAO, *Crop Insurance: Savings Would Result from Program Changes and Greater Use of Data Mining,* GAO-12-256 (Washington, D.C.: Mar. 13, 2012).

[9] GAO, *Climate Change: Financial Risks to Federal and Private Insurers in Coming Decades Are Potentially Significant*, GAO-07-285 (Washington, D.C.: Mar. 16, 2007).

[10] To assess the reliability of exposure data, we reviewed federal agency procedures for collecting and updating the data. For private sector exposure data, we compared exposure estimates from two separate firms. To assess the reliability of scientific studies, staff from our Chief Scientist's office and Center for Economics reviewed each study to determine whether their methodologies were sound and appropriate for our purposes.

[11] Because we used a nonprobability sample to select states, our results are not generalizable to all 50 states; however, our results can provide illustrative information.

[12] According to National Association of Insurance Commissioners (NAIC) data, property and casualty insurers wrote around $521.87 billion in premiums in the United States in 2012, with 11 insurance companies comprising about 50 percent of the property and casualty market. Also in 2012, reinsurers wrote about $32.98 billion in reinsurance premiums in the United States, a 10.5 percent increase from 2011.

[13] Melillo, Richmond, and Yohe, eds., *Climate Change Impacts in the United States: The Third National Climate Assessment*, and Field, C.B. et al., eds., Summary for Policymakers in:

Managing the Risks of Extreme Events and Disasters to Advance Climate Change Adaptation, A Special Report of Working Groups I and II of the Intergovernmental Panel on Climate Change (Cambridge University Press, Cambridge, UK, and New York, NY, USA, 2012).

[14] NFIP regulations identify minimum requirements that communities must fulfill to join and stay in the program. The requirements that apply to a particular community depend on its flood hazard and the level of detail of the data FEMA provides to the community.

[15] Any property owner in a participating community may purchase flood insurance. However, in participating communities, certain property owners are required to purchase flood insurance. Specifically, owners of properties in special flood hazard (SFHA) areas who obtain mortgages from federally regulated lending institutions or federal agency lenders or who receive direct financial assistance for acquisition or construction of the property are required to purchase flood insurance. In addition, individuals in SFHAs who receive federal disaster assistance after September 23, 1994, for flood losses to real or personal property are also required to purchase and maintain flood insurance on the property as a condition for receiving future disaster assistance. FEMA defines SFHAs as land in the flood plain subject to a 1 percent or greater chance of flooding in any given year—also referred to as the "100-year flood."

[16] Depending on the level of activities that communities undertake in four areas—public information, mapping and regulatory activities, flood damage reduction, and flood preparedness—communities fall into 1 of 10 rate reduction classes. See GAO-14-297R.

[17] For example, as mentioned previously, recently enacted legislation required FEMA to phase out subsidies for some properties, such as nonprimary residences and business structures, through 25 percent annual premium increases until the full-risk rate is reached.

[18] For more information, see GAO, *Fiscal Exposures: Improving the Budgetary Focus on Long-Term Costs and Uncertainties*, GAO-03-213 (Washington, D.C.: Jan. 24, 2003); and GAO-14-28.

[19] Land subsidence is the gradual settling or sudden sinking of the earth's surface, whereas uplift is the rising of the earth's surface caused by geophysical forces. Both land subsidence and uplift can affect sea-level rise.

[20] According to NOAA officials, for shorter time frames, natural climate variability can limit predictability.

[21] NRC, America's Climate Choices: Panel on Adapting to the Impacts of Climate Change, *Adapting to the Impacts of Climate Change* (Washington, D.C.: 2010). NRC is the principal operating agency of the National Academy of Sciences and the National Academy of Engineering.

[22] GAO, *Disaster Resilience: Actions Are Underway, but Federal Fiscal Exposure Highlights the Need for Continued Attention to Longstanding Challenges*, GAO-14-603T (Washington, D.C.: May 14, 2014).

[23] Executive Order 13653: *Preparing the United States for the Impacts of Climate Change,* November 6, 2013.

[24] Private sector exposure is based on estimates from two insurance industry modeling firms, AIR Worldwide and Risk Management Solutions (RMS). Growth is shown through 2012 rather than 2013 because RMS and AIR Worldwide did not calculate estimates for 2013.

[25] For the purposes of this report, we included data on exposure during the time frame covered by our prior work for context.

[26] We compared the median value of disaster appropriations over a multiyear time period to control for large fluctuations in annual disaster relief appropriations from year to year.

[27] Congressional Research Service, *Disaster Relief Funding and Supplemental Appropriations for Disaster Relief*, R40708 (Washington, D.C.: Aug. 5, 2013) and Congressional Research Service, *FEMA's Disaster Relief Fund: Overview and Selected Issues*, R43537 (Washington, D.C.: May 7, 2014). Not all disasters are weather-related, such as earthquakes.

[28] J. David Cummins, Michael Suher, and George Zanjani, "Federal Financial Exposure to Natural Catastrophe Risk," chapter in *Measuring and Managing Federal Financial Risk*, National Bureau of Economic Research, ed. Deborah Lucas (Chicago: University of Chicago Press, 2010) 61-92 and GAO-14-28.

[29] Some of the studies we reviewed did find a statistically significant trend in losses for certain weather events, however, studies also cited a variety of potential reasons other than climate change for the identified trends, such as natural variability.

[30] Summary for Policymakers in: *Managing the Risks of Extreme Events and Disasters to Advance Climate Change Adaptation,* A Special Report of Working Groups I and II of the Intergovernmental Panel on Climate Change (Cambridge University Press, Cambridge, UK, and New York, NY, USA, 2012).

[31] McAneney, J., et al., *Market-Based Mechanisms for Climate Change Adaptation: Assessing the Potential for and Limits to Insurance and Market-Based Mechanisms for Encouraging Climate Change Adaptation,* National Climate Change Adaptation Research Facility (Gold Coast, Australia: 2013) and Mendelsohn et al., *The Impact of Climate Change on Global Tropical Storm Damages,* The World Bank Policy Research Working Paper (Washington, D.C.: 2011).

[32] Mendelsohn et al.,"The Impact of Climate Change on Global Tropical Cyclone Damage." *Nature Climate Change*, vol. 2 (2012) 205-209.

[33] Melillo, Richmond, and Yohe, eds., *Climate Change Impacts in the United States: The Third National Climate Assessment.*

[34] AECOM, *The Impact of Climate Change and Population Growth on the National Flood Insurance Program Through 2100*, a report prepared for the Federal Insurance and Mitigation Administration, Federal Emergency Management Agency (June 2013).

[35] Pub. L. No. 112-141, div. F, tit. II, subtit. A, §100215 126 Stat. 405, 924-27 (2012).

[36] RTI International, *Climate Change Impacts on Crop Insurance*, a report prepared at the request of Risk Management Agency, U.S. Department of Agriculture (December 2009).

[37] The PRISM Climate Group gathers climate observations from a wide range of monitoring networks, puts its data and analysis through peer reviews, and develops spatial climate datasets to reveal short- and long-term climate patterns. The resulting datasets incorporate a variety of modeling techniques and are available at multiple spatial/temporal resolutions, covering the period from 1895 to the present.

[38] Executive Order 13514, *Federal Leadership in Environmental, Energy, and Economic Performance* and Executive Order 13653, *Preparing the United States for the Impacts of Climate Change*.

[39] Section 508(i)(2) of the Federal Crop Insurance Act as amended requires RMA to periodically review the methodologies for rating plans.

[40] RMA can still use historical data more than 20 years old for estimating potential catastrophic events. However, in the total rate calculation, the old data will receive less weight than in pre-2011 calculations.

[41] This task force was established by Executive Order 13632: *Establishing the Hurricane Sandy Rebuilding Task Force,* December 7, 2012.

[42] According to a July 2014 White House statement, FEMA will release new guidance for state hazard mitigation plans that calls upon states to consider climate variability as part of their requirement to address the probability of future events in state planning efforts, in an effort to rebuild stronger and safer after natural disasters and ensure that states prepare for the impacts of climate change. The White House, Office of the Press Secretary, July 16, 2014, *Fact Sheet: Taking Action to Support State, Local, and Tribal Leaders as They Prepare Communities for the Impacts of Climate Change,* accessed August 1, 2014, http://www.whitehouse.gov/the-press-office/2014/07/16/fact-sheet-taking-action-support-state-local-and-tribal-leaders-they-pre.

[43] FEMA's Hazard Mitigation Assistance programs include the Hazard Mitigation Grant Program, which assists in implementing long-term hazard mitigation measures following a major disaster; the Pre-Disaster Mitigation Program, which provides funds for hazard mitigation planning and projects on an annual basis, and Flood Mitigation Assistance Program, which provides funds for projects to reduce risk of flood damage to buildings that are insured under NFIP on an annual basis.

[44] For reinsurance and catastrophe bonds, the market sets the cost. The transaction occurs at a price acceptable to both the buyer(s) and seller(s) of the risk.

[45] GAO, *Flood Insurance: Strategies for Increasing Private Sector Involvement,* GAO-14-127 (Washington, D.C.: Jan. 22, 2014). NFIP's operating environment differs from that of traditional insurers and limits FEMA's ability to keep the program financially sound. In particular, NFIP assumes and retains all of the risks for the policies it sells, is required to accept virtually all applicants for insurance, and cannot deny coverage for potentially high-risk properties.

[46] Although the Homeowners Flood Insurance Affordability Act requires FEMA to phase out premium subsidies, this process may take several years. Consequently, subsidized rates remain a program challenge.

[47] We previously found that subsidized policies in 2012 accounted for about 20 percent of NFIP policies, and premium subsidies accounted for about 62 percent of crop insurance premiums collected by RMA in 2011. See GAO, *Flood Insurance: More Information Needed on Subsidized Properties*, GAO-13-607 (Washington, D.C.: July 3, 2014) and GAO-12-256.

[48] According to RMA, under the crop insurance program, insurance providers notify every producer about how much premium was paid by the government and how much is paid by the producer. Under the Homeowner Flood Insurance Affordability Act of 2014, FEMA is required to communicate full-risk flood determinations to individual property owners regardless of whether their rates are full-risk or subsidized. However, in a July 2013 report, we found that property elevations—which inform flood risk determinations—are unknown for 97 percent of both 1.15 million historically subsidized policies and more than 700,000 subsidized policies in special flood hazard areas. In the report, we concluded that FEMA will need elevation certificates to phase out and eventually eliminate subsidies and revise rates over time. GAO-13-607. We recommended that FEMA develop and implement a plan, including a timeline, to obtain needed elevation information as soon as practicable.

[49] Pub. L. No. 112-141, div. F, tit.ll, subtit. A (2012), Homeowner Flood Insurance Affordability Act of 2014, Pub. L. No. 113-89 (2014).

[50] According to FEMA guidance, floodproofing is any combination of structural and nonstructural additions, changes, or adjustments to structures which reduce or eliminate flood damage to real estate or improved real property, water and sanitary facilities, structures and their contents. For example, a property owner could make their structure watertight below base flood elevation.

[51] See also GAO-12-256 for more information on historical premium subsidy rates.

[52] GAO, *Crop Insurance: Considerations in Reducing Federal Premium Subsidies,* GAO-14-700 (Washington, D.C.: Aug. 8, 2014).

[53] For example, if a farmer has to make a decision about whether to farm land that may pose a higher risk of loss, the farmer is less likely to do so if he or she would have to pay a full rather than a subsidized insurance premium.

[54] GAO-14-700, GAO-13-607, and GAO-12-256. We have also found that the federal government's cash-based budgeting does not adequately reflect the government's cost or the economic impact of federal insurance programs. Cash-based budgeting generally recognizes costs as they are paid, whereas accrual-based budgets recognize costs when commitments are made. We concluded the potential benefits of an accrual-based budgeting approach for federal insurance programs warrant continued effort in the development of risk-assumed cost estimates. See GAO/T-AIMD-98-147. As we previously found, the full cost of crop insurance subsidies and administrative costs is reflected in the budget. GAO-14-28.

[55] If insurance contracts were long-term, other, different challenges could result. For example, long-term contracts could be more difficult to price and could be prohibitively expensive because of the need for greater risk margins reflecting the greater uncertainty of the magnitude of long-term changes.

[56] For example, RMA officials said that appropriate farming practices vary depending on the farm location and the conditions specific to any given year.

[57] According to NAIC, the climate risk disclosure survey was designed to inform state regulators about how insurers incorporate climate change into their risk management practices and provide benchmarks that insurers could use to assess their climate change strategies and identify how climate change affects their business. The survey consisted of eight questions, asking insurers to describe how they incorporate climate risks into their mitigation, risk management, and investment plans, and to identify steps taken to engage key constituencies and policyholders on the topic of climate change.

[58] The California Department of Insurance presented its analysis of the 2012 survey results in March 2014 to NAIC's climate change working group. We did not assess the reliability of the California Department of Insurance's findings.

[59] Two catastrophe modeling firms whose representatives we interviewed have developed hurricane models that include both historical weather information and projections based on sea surface temperature changes, but most catastrophe models are based on historical weather data and scenario-modeling based on previous weather events.

[60] McChristian, Lynne, *Hurricane Andrew and Insurance: The Enduring Impact of an Historic Storm* (Tampa, FL: Insurance Information Institute, August 2012).

[61] Some insurers offer incentives to customers for mitigation. For example, one insurer we interviewed offers discounted premiums on homeowners' insurance to homeowners that have taken steps to improve their structures ability to withstand strong winds and hail. In addition, Florida law requires insurers licensed in the state to offer discounts to homeowners who take certain steps to mitigate wind storm damage. Further, several industry representatives said insurers provide educational information to homeowners to encourage both mitigation and preparedness for risks associated with climate change such as hurricanes, floods, and wildfires.

[62] South Carolina's Omnibus Coastal Property Insurance Reform Act of 2007 provides state income tax credits for the costs a homeowner incurs to retrofit one's home to make it more

resistant to loss due to hurricanes, rising floodwaters, or other catastrophic windstorm events. S.C. Code Ann. § 12-6-3660.

[63] As one industry group has reported, several states have not updated their building codes to reflect the most recent international standards. As such, these codes may not protect structures against current risk, regardless of climate change. See Insurance Institute for Business and Home Safety, *Rating the States: An Assessment of Residential Building Code and Enforcement Systems for Life Safety and Property Protection in Hurricane-Prone Regions* (Tampa, FL: Insurance Institute for Business and Home Safety, December 2011).

[64] Hurricane Sandy Rebuilding Task Force, *Hurricane Sandy Rebuilding Strategy: Stronger Communities, a Resilient Region* (Hurricane Sandy Rebuilding Task Force, August 2013).

[65] Insurance Institute for Business and Home Safety, *Rating the States* (December 2011).

[66] According to White House press releases, in March 2014, the White House launched a Climate Data Initiative to leverage government resources and data to build tools that will make communities more resilient, in partnership with the private sector. In addition, in June 2014, senior administration officials met with a group of insurance and reinsurance industry leaders to discuss the economic consequences of increasingly frequent and severe extreme weather. The discussion centered on opportunities to share data between the federal government and the insurance industry to better communicate and reduce risks to policyholders, communities, and taxpayers. Participants also agreed to find ways to collaborate to create broader community resilience to strengthen neighborhoods and businesses.

[67] American Institutes for Research, *Assessing the Adequacy of the National Flood Insurance Program's 1 Percent Flood Standard*, a report prepared under subcontract to the American Institutes for Research as part of the 2001-2006 Evaluation of the National Flood Insurance Program (October, 2006); and American State Floodplain Managers Foundation, *Reducing Flood Losses: Is the 1% Chance Flood Standard Sufficient?*, a report of the 2004 Assembly of the Gilbert F. White National Flood Policy Forum (September, 2004).

End Note for Appendix I

[1] To assess the reliability of exposure data, we reviewed federal agency procedures for collecting and updating the data. For private sector exposure data, we compared exposure estimates from two separate firms. To assess the reliability of scientific studies, staff from our Center for Science, Technology and Engineering office and Center for Economics reviewed each study to determine whether their methodologies were sound and appropriate for our purposes.

In: Climate Change's Effect on Insurers
Editor: Noah Lynch
ISBN: 978-1-63482-518-4

Chapter 2

CLIMATE CHANGE: FINANCIAL RISKS TO FEDERAL AND PRIVATE INSURERS IN COMING DECADES ARE POTENTIALLY SIGNIFICANT*

United States Government Accountability Office

WHY GAO DID THIS STUDY

Weather-related events have cost the nation billions of dollars in damages over the past decade. Many of these losses are borne by private insurers and by two federal insurance programs—the National Flood Insurance Program (NFIP), which insures properties against flooding, and the Federal Crop Insurance Corporation (FCIC), which insures crops against drought or other weather disasters.

GAO was asked to (1) describe how climate change may affect future weather-related losses, (2) determine past insured weather-related losses, and (3) determine what major private insurers and federal insurers are doing to prepare for potential increases in such losses. In response, among other things, GAO reviewed key scientific assessments; analyzed insured loss data; and contacted private insurers, NFIP, and FCIC.

* This is an edited, reformatted and augmented version of a United States Government Accountability Office publication, No. GAO-07-285, dated March 2007.

WHAT GAO RECOMMENDS

GAO is recommending that the Secretaries of Agriculture and Homeland Security analyze the potential long-term fiscal implications of climate change for the FCIC and the NFIP, respectively, and report their findings to the Congress. In commenting on a draft of this report, the two agencies agreed with the recommendation. The Departments of Agriculture and Commerce made comments and suggestions on the presentation of several findings. The Department of Energy elected not to comment.

WHAT GAO FOUND

Key scientific assessments report that the effects of climate change on weather-related events and, subsequently, insured and uninsured losses, could be significant. The global average surface temperature has increased by 0.74 degrees Celsius over the past 100 years and climate models predict additional, perhaps accelerating, increases in temperature. The key assessments GAO reviewed generally found that rising temperatures are expected to increase the frequency and severity of damaging weather-related events, such as flooding or drought, although the timing and magnitude are as yet undetermined. Additional research on the effect of increasing temperatures on weather events is expected in the near future, including a highly anticipated assessment of the state of climate science this year.

Taken together, private and federal insurers paid more than $320 billion in claims on weather-related losses from 1980 to 2005. Claims varied significantly from year to year—largely due to the effects of catastrophic weather events such as hurricanes and droughts—but have generally increased during this period. The growth in population in hazard-prone areas and resulting real estate development have generally increased liabilities for insurers, and have helped to explain the increase in losses. Due to these and other factors, federal insurers' exposure has grown substantially. Since 1980, NFIP's exposure quadrupled, nearing $1 trillion in 2005, and program expansion increased FCIC's exposure 26-fold to $44 billion.

Major private and federal insurers are both exposed to the effects of climate change over coming decades, but are responding differently. Many large private insurers are incorporating climate change into their annual risk management practices, and some are addressing it strategically by assessing its potential long-term industry-wide impacts. The two major federal insurance programs, however, have done little to develop comparable information. GAO acknowledges that the federal insurance programs are not profit-oriented, like private insurers. Nonetheless, a strategic analysis of the potential implications of climate change for the major federal insurance programs would help the Congress manage an emerging high-risk area with significant implications for the nation's growing fiscal imbalance.

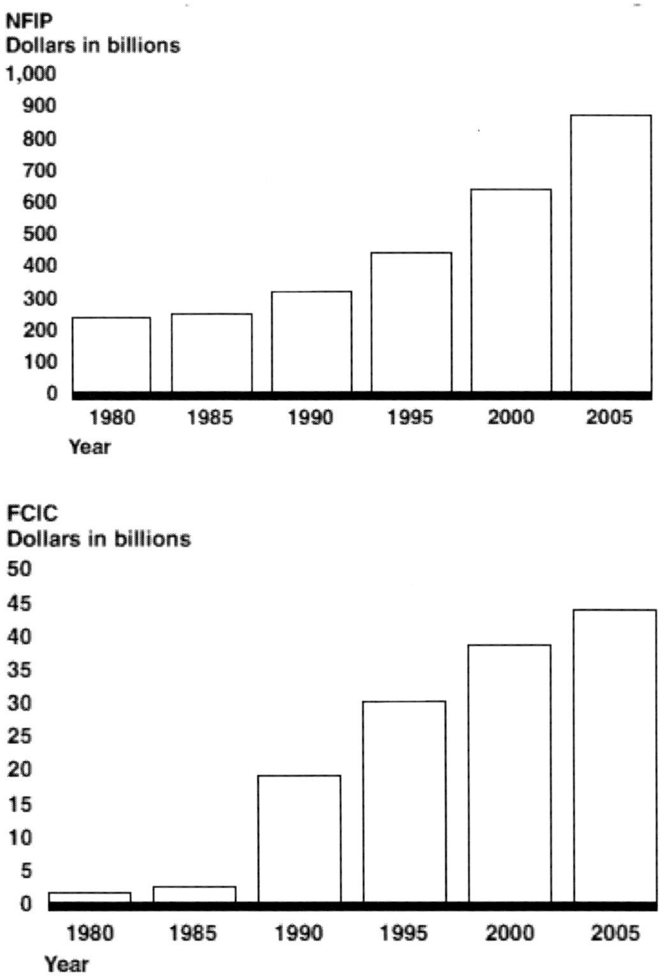

Source: GAO.

Growth in Exposure of Federal Insurance Programs ($2005).

ABBREVIATIONS

AAA	American Academy of Actuaries
AMO	Atlantic Multidecadal Oscillation
CCSP	Climate Change Science Program
FAIR	Fair Access to Insurance Requirements
FEMA	Federal Emergency Management Agency
FCIC	Federal Crop Insurance Corporation
HUD	Department of Housing and Urban Development
IPCC	Intergovernmental Panel on Climate Change
NAIC	National Association of Insurance Commissioners
NAS	National Academy of Sciences
NFIP	National Flood Insurance Program
NHC	National Hurricane Center
NOAA	National Oceanic and Atmospheric Administration
PCS	Property Claim Services
RMA	risk Management Agency
SAP	synthesis and assessment product
SFIP	standard flood insurance policy
USDA	U.S. Department of Agriculture

* * *

March 16, 2007

The Honorable Joseph I. Lieberman
Chairman

The Honorable Susan M. Collins
Ranking Member
Committee on Homeland Security and Governmental Affairs
United States Senate

As the 2004 and 2005 hurricane seasons demonstrated, weather-related events can devastate affected communities and individuals, and are costly to the insurance industry, government disaster assistance programs, and other relief organizations. Apart from the record-setting losses experienced in 2005, weather-related events over the past decade have cost the country tens of billions of dollars each year.

The property and casualty segment of the insurance industry, spanning both the private and public sector, bears a large portion of weather-related losses.[1] The private sector includes primary insurers that insure individuals and businesses directly, and reinsurers that provide insurance to the primary insurers. The public sector includes federal programs—in particular, the National Flood Insurance Program (NFIP), which insures properties at risk of damage from flooding, and the Federal Crop Insurance Corporation (FCIC), which insures crops that are vulnerable to drought, floods, or other natural disasters. Many states also administer insurance pools that provide coverage for losses caused by weather-related events.

The uncertain and potentially large losses associated with weather-related events are among the biggest risks that property insurers face. Virtually anything that is insured—property, crops and livestock, business operations, or human life and health—is vulnerable to weather-related events. To remain financially solvent, the insurance industry must estimate and prepare for the potential impact of weather-related events. As such, any unanticipated changes in the frequency or severity of weather-related events can have financial consequences at the company level and industry-wide.

The earth's climate and weather patterns are dynamic, varying on seasonal, decadal, and longer time scales. The global average surface temperature has increased by 0..74 degrees Celsius over the past 100 years and climate models predict additional, perhaps accelerating, increases in temperature. While the temperature increases to date may appear small, climate models project that additional changes in temperature may alter social and economic activities in potentially profound ways. Much research and policy debate has centered on the extent to which human activities have contributed to the warming and how much is due to natural variability. For the purposes of this report, *climate change* refers to any change in the climate over time, whether due to natural variability or as a result of human activity.[2] Regardless of the cause, some contend that increasing temperatures—accompanied by changes in other aspects of the climate—may have adverse financial consequences for property insurers, which might slow the growth of the industry and shift more of the burden to governments and individuals.

Concerned about the implications of climate change for weather-related losses incurred by federal agencies and private insurers, you asked us to (1) describe what is known about how climate change might affect insured and uninsured losses, (2) determine insured losses incurred by major federal agencies and private insurers and reinsurers resulting from weather-related events, and (3) determine what major federal agencies and private insurers and reinsurers are doing to prepare for the potential risk of increased losses due to more frequent or more severe weather-related events associated with climate change.

To describe how climate change might affect insured and uninsured losses, we reviewed and summarized key scientific assessments by reputable international and national research organizations, including the Intergovernmental Panel on Climate Change *Third Assessment Report*, National Academy of Sciences reports, and the multifederal agency Climate Change Science Program. To determine insured losses attributable to weather-related events, we analyzed data from 1980 through 2005 from the Department of Homeland Security's Federal Emergency Management Agency (FEMA) for the NFIP; from the Department of Agriculture's Risk Management Agency (RMA) for FCIC; and from the Property Claims Service, a leading source of insurance data. We analyzed changes in weather-related losses since 1980 and supplemented this analysis with a review of existing literature and the views of subject area experts on the key drivers of changes in losses.

To determine what key federal agencies and private insurers are doing to assess and manage the potential for increased losses, we conducted semistructured interviews with officials from the NFIP, RMA, and a sample of the largest private primary insurers and reinsurers in the United States, Europe, and Bermuda. The companies we interviewed represent about 45 percent of the total domestic insurance market but should not be generalized to represent all insurance companies. We also interviewed officials from catastrophe modeling firms, insurance industry associations, the National Association of Insurance Commissioners (NAIC),[3] and universities to provide additional context for respondents' statements. To supplement these interviews, we reviewed documentation of federal agencies' risk management practices, studies by subject area experts, industry reports, insurance company documents, and previous GAO reports. We performed our work between February 2006 and January 2007 in accordance with generally accepted government auditing standards. A more extensive discussion of our scope and methodology appears in appendix I.

RESULTS IN BRIEF

Assessments by the National Academy of Sciences (NAS) and the Intergovernmental Panel on Climate Change (IPCC), a leading source for international climate expertise, report that the effects of climate change on weather-related events and—by extension—weather-related losses could be substantial. IPCC reports that global mean temperatures increased by 0.74 degrees Celsius over the last 100 years and are projected to continue to rise over the next century. Although temperatures have varied throughout history due to natural processes, such as changes in the Earth's orbit and volcanic eruptions, the IPCC and NAS report that the observed temperature increase during the twentieth century cannot be explained by natural variability alone but is largely attributable to human activities. Warmer surface temperatures are linked to global-scale oceanographic, meteorological, and biological changes. For example, as the earth warms, more water evaporates from oceans and other sources, eventually falling as rain or snow. Key assessments that rely on both observational data and computer models have reported that warmer temperatures are expected to increase the frequency and severity of damaging extreme weather-related events (such as flooding or drought), although the timing, magnitude, and duration of these changes are as yet undetermined. Further research on the effect of increasing temperature on weather events is ongoing. Of particular note, the IPCC is expected to release its fourth assessment of the state of climate science throughout 2007, and the Climate Change Science Program is currently assessing potential changes in the frequency or intensity of weather-related events specific to North America in a report scheduled for release in 2008.

Taken together, private and federal insurers paid more than $320 billion in claims on weather-related losses from 1980 through 2005. In constant dollars, private insurers paid the largest part of the claims during this period, $243.5 billion (about 76 percent); followed by federal crop insurance, $43.6 billion (about 14 percent); and federal flood insurance, $34.1 billion (about 11 percent). Claims varied significantly from year to year—largely due to the incidence and effects of catastrophic weather events such as hurricanes and droughts—but generally increased during this period. In particular, the years with the largest insured losses were generally associated with major hurricanes, which comprised well over one-third of all weather-related losses since 1980. The growth in population in hazard-prone areas, and resulting real estate development and increasing real estate values, have increased federal and private insurers' exposure, and have helped to explain the increase in losses. In particular, heavily-populated areas along the Northeast, Southeast, and Texas coasts have among the highest value of insured properties in the United States and face the highest likelihood of major hurricanes. Due to these and other factors, federal insurers' exposures have grown substantially. Since 1980, NFIP's exposure has quadrupled, nearing $1 trillion, and program expansion has increased FCIC's exposure nearly 26- fold to $44 billion. These escalating exposures to catastrophic weather events are leaving the federal government at increased financial risk. FCIC officials told us, for example, that if the widespread Midwest floods of 1993 were to occur today, losses would be five times greater.

While both major private and federal insurers are exposed to increases in the frequency or severity of weather-related events associated with climate change, the two sectors are responding in different ways. Using computer-based catastrophe models, many major private insurers are incorporating some near-term elements of climate change into their risk management practices. One consequence is that, as these insurers seek to limit their own catastrophic risk exposure, they are transferring some of it to policyholders and to the public sector. In addition, some private insurers are approaching climate change at a strategic level by publishing reports outlining the potential industry-wide impacts and strategies to proactively address the issue. Federal insurance programs, on the other hand, have done little to develop the kind of information needed to understand the programs' long-term exposure to climate change for a variety of reasons. The federal insurance programs are not oriented toward earning profits like private insurers but rather toward increasing participation among eligible parties. Consequently, neither program has had reason to develop information on their long-term exposure to the fiscal risks associated with climate change.

We acknowledge the different mandate and operating environment in which the major federal insurance programs operate, but we believe that better information about the federal government's exposure to potential changes in weather-related risk would help the Congress identify and manage this emerging high-risk area—one which may not constitute an immediate crisis, but which does have significant implications for the nation's growing fiscal imbalance. Accordingly, GAO is recommending that the Secretary of Agriculture and the Secretary of Homeland Security direct the Under Secretary for Farm and Foreign Agricultural Services and the Under Secretary of Homeland Security for Emergency Preparedness to analyze the potential long-term fiscal implications of climate change for the FCIC and the NFIP, respectively, and report their findings to the Congress.

In commenting on a draft of this report, both the Departments of Agriculture (USDA) and Homeland Security (DHS) agreed with our recommendation, and USDA commented on the presentation of several findings in the draft. The Department of Commerce neither agreed nor disagreed with the report's findings, but instead commented on the presentation of several issues in the draft and offered technical comments which we incorporated into this report as appropriate. The Department of Energy elected not to provide comments on the draft.

BACKGROUND

Insurance is a mechanism for spreading risk over time, across large geographical areas, and among industries and individuals. While insurers assume some financial risk when they write policies, they employ various strategies to manage risk so that they earn profits, limit potential financial exposures, and build capital needed to pay claims.[4] For example, they charge premiums for coverage and establish underwriting standards, such as refusing to insure customers who pose unacceptable levels of risk, or limiting coverage in particular geographic areas. Insurance companies may also purchase reinsurance to cover specific portions of their financial risk. Reinsurers use similar strategies to limit their risks, including charging premiums, establishing underwriting standards, and maintaining close, long-term business relationships with certain insurers.

Both insurers and reinsurers must also predict the frequency and severity of insured losses with some reliability to best manage financial risk.[5] In some cases, these losses may be fairly predictable. For example, the incidence of most automobile insurance claims is predictable, and losses generally do not occur to large numbers of policyholders at the same time. However, some infrequent weather-related events—hurricanes, for example—are so severe that they pose unique challenges for insurers and reinsurers. Commonly referred to as catastrophic or extreme events, the unpredictability and sheer size of these events—both in terms of geography and number of insured parties affected—have the potential to overwhelm insurers' and reinsurers' capacity to pay claims. Catastrophic events may affect many households, businesses, and public infrastructure across large areas, resulting in substantial losses that deplete insurers' and reinsurers' capital.

Given the higher levels of capital that reinsurers must hold to address catastrophic events, reinsurers generally charge higher premiums and restrict coverage for such events. Further, in the wake of catastrophic events, reinsurers and insurers may sharply increase premiums to rebuild capital reserves and may significantly restrict insurance and reinsurance coverage to limit exposure to similar events in the future.

Under certain circumstances, the private sector may determine that a risk is uninsurable. For example, while homeowner insurance policies typically cover damage and losses from fire and other perils, they usually do not cover flood damage because private insurance companies are largely unwilling to bear the financial risks associated with its potentially catastrophic impact. In other instances, the private sector may be willing to insure a risk, but at rates that are not affordable to many property owners. Without insurance, affected property owners must rely on their own resources or seek out disaster assistance from local, state, and federal sources.

In situations where the private sector will not insure a particular type of risk, the public sector may create markets to ensure the availability of insurance. For example, several states have established Fair Access to Insurance Requirements (FAIR) plans, which pool resources from insurers doing business in the state to make property insurance available to property owners who cannot obtain coverage in the private insurance market, or cannot do so at an affordable rate. In addition, six southern states have established windstorm insurance pools that pool resources from private insurers to make insurance available to property owners who cannot obtain it in the private insurance market.

Similarly, at the federal level, the Congress established the NFIP and the FCIC to provide coverage where voluntary markets do not exist.[6] The Congress established the NFIP in 1968, partly to provide an alternative to disaster assistance for flood damage. Participating communities are required to adopt and enforce floodplain management regulations, thereby reducing the risks of flooding and the costs of repairing flood damage. FEMA, within the Department of Homeland Security, is responsible for, among other things, oversight and management of the NFIP. Under the program, the federal government assumes the liability for covered losses and sets rates and coverage limitations.

The Congress established the FCIC in 1938 to temper the economic impact of the Great Depression and the weather effects of the dust bowl. In 1980, the Congress expanded the program to provide an alternative to disaster assistance for farmers that suffer financial losses when crops are damaged by droughts, floods, or other natural disasters. Farmers' participation is voluntary, but the federal government encourages it by subsidizing their insurance premiums. USDA's RMA is responsible for administering the crop insurance program, including issuing new insurance products and expanding existing insurance products to new geographic regions. RMA administers the program in partnership with private insurance companies, which share a percentage of the risk of loss or the opportunity for gain associated with each insurance policy written.

CLIMATE CHANGE MAY INCREASE LOSSES BY ALTERING THE FREQUENCY OR SEVERITY OF WEATHER-RELATED EVENTS

Global temperatures have increased in the last 100 years and are projected to continue to rise over the next century. Using observational data and computer modeling, climatologists and other scientists are assessing the likely effects of temperature rise associated with climate change on precipitation patterns and on the frequency and severity of weather-related events. The key scientific assessments we reviewed generally found that warmer temperatures are expected to alter the frequency or severity of damaging weather-related events, such as flooding or drought, although the timing, magnitude, and duration of these changes are as yet undetermined. Additional research on the effect of increasing temperature on weather events is expected in the near future. Nevertheless, research suggests that the potential effects of climate change on damaging weather-related events could be significant.

Warming Temperatures Are Expected to Alter the Frequency and Severity of Damaging Extreme Weather-Related Events

We reviewed the reports released by IPCC, NAS, and the federal Climate Change Science Program (CCSP) that are shown in figure 1.[7] These leading scientific bodies report that the Earth warmed during the twentieth century—0.74 degrees Celsius from 1906 to 2005 according to a recent IPCC report—and is projected to continue to warm for the foreseeable future.[8] IPCC, NAS, CCSP, and other scientific bodies report that this increase in temperature cannot be explained by natural variation alone. IPCC's 2001 assessment of the impact of increasing temperatures on extreme weather events found that it was likely the frequency and severity of several types of events will increase as greenhouse gas emissions continue.[9]

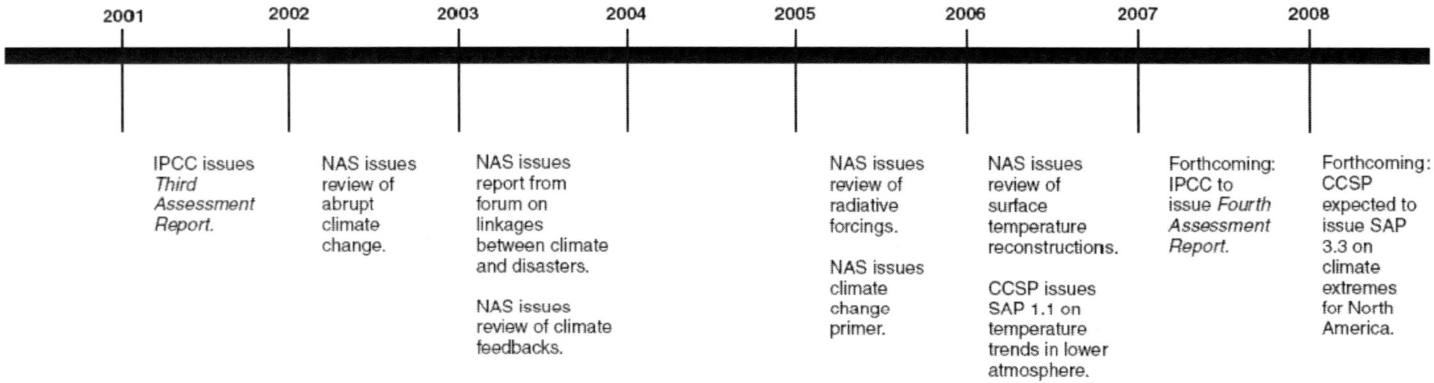

Source: GAO.

Figure 1. Time Line of Key Scientific Assessments.

Average Global Temperatures Have Increased and Are Expected to Continue to Rise

The earth's climate system is driven by energy from the sun and is maintained by complex interactions between the atmosphere, the oceans, and the reflectivity of the earth's surface, among other factors. Upon reaching the earth, the sun's energy is either reflected back into space, or is absorbed by the earth and is subsequently reemitted. However, certain gases in the earth's atmosphere—such as carbon dioxide and methane— act like the glass in a greenhouse to trap some of the sun's energy and prevent it from returning to space. While these gases play an important part in maintaining life on earth, their accumulation in the atmosphere can significantly increase global temperatures.

The earth warmed by roughly 0.74 degrees Celsius over the past 100 years, and is projected to continue warming for the foreseeable future. While temperatures have varied throughout history, triggered by natural factors such as volcanic eruptions or changes in the earth's orbit, the key scientific assessments we reviewed have generally concluded that the observed increase in temperature in the past 100 years cannot be explained by natural variability alone. In recent years, major scientific bodies such as the IPCC, NAS, and the Royal Academy (the United Kingdom's national academy of science) have concluded that human activities, including the combustion of fossil fuels, industrial and agriculture processes, landfills, and some land use changes, are significantly increasing the concentrations of greenhouse gases and, in turn, global temperatures.

Although climate models produce varying estimates of the extent of future changes in temperature, NAS and other scientific organizations have concluded that available evidence points toward continued global temperature rise. Assuming continued growth in atmospheric concentration of greenhouse gases, the latest assessment of computer climate models projects that average global temperatures will warm by an additional 1.8 to 4.0 degrees Celsius during the next century.[10]

Some scientists have questioned the significance of the earth's present temperature rise relative to past fluctuations. To address this issue, the NAS recently assessed the scientific community's efforts to reconstruct temperatures of the past 2,000 years and place the earth's current warming in an historical context.[11] Based on its review, the NAS concluded with a high level of confidence that global mean surface temperature was warmer during the last few decades of the twentieth century than during any comparable period during the preceding 400 years. Moreover, NAS cited evidence that temperatures at many, but not all, individual locations were higher during the past 25 years than any period of comparable length over the past 1,100 years.

IPCC Expects Continued Warming to Alter Frequency and Severity of Damaging Extreme Weather-Related Events

Determining the precise nature and extent of the relationship between average global temperatures and weather-related events is an exceedingly challenging task. Several key assessments of the state of this science have addressed the large body of work on this topic. Using observational data and computer models, scientists are examining the effects of rising temperatures on precipitation patterns and the frequency and severity of extreme weather-related events. The complexity of weather systems, together with the limited statistical precision of projections of the extent of future temperature change, often produces different model results, and the results themselves represent a range of potential future conditions.

Nonetheless, a key assessment of climate model projections indicates that an increase is likely in the frequency or severity of damaging extreme weather-related events. In 2001, the IPCC, a leading scientific authority on climate science, released its *Third Assessment Report*, which assessed the state of knowledge of, among other things, the potential for global changes in extreme weather-related events. The IPCC described the relationship between temperatures, precipitation, and weather-related events. Increased global mean surface temperatures are linked to global-scale oceanographic, meteorological, and biological changes. For example, as the earth warms, more water evaporates from oceans or lakes, eventually falling as rain or snow. IPCC reported that permafrost is thawing, and the extent of sea ice, snow cover, and mountain glaciers are generally shrinking. The IPCC also noted that global sea level rose between 0.1 and 0.2 meters during the twentieth century through thermal expansion of seawater and widespread loss of land ice, and that this sea level rise could increase the magnitude of hurricane storm surge in some areas. Warming is expected to change rainfall patterns, partly because warmer air holds more moisture.

Based on model projections and expert judgment,[12] the IPCC reported that future increases in the earth's temperature are likely to increase the frequency and severity of many damaging extreme weather-related events (summarized in table 1). For instance, IPCC reported that increased drought is likely across many regions of the globe, including the U.S. Great Plains. Also, IPCC concluded that the intensity of precipitation events is very likely to increase across almost all regions of the globe and that heavy precipitation events are expected to become more frequent. Compared with projected temperature increases, changes in the frequency and severity of extreme events can occur relatively rapidly, according to the IPCC.

Table 1. Selected IPCC Estimates of Confidence in Projected Changes in Weather-Related Events

Weather-related event	Confidence in projected future changes
Higher maximum temperatures and more hot days over nearly all land areas	Very likely
Higher minimum temperatures and fewer cold and frost days over nearly all land areas	Very likely
More intense precipitation events	Very likely
Increased summer drying and associated risks of drought	Likelya
Increase in hurricane peak wind intensities	Likelyb
Increase in hurricane average and peak precipitation intensities	Likely

Source: IPCC, Climate Change 2001: The Scientific Basis, 2001.

[a]Projections for most midlatitude continental interiors. IPCC found a lack of consistent projections in other regions.

[b]IPCC reported that changes in the regional distribution of hurricanes are possible but have not been established.

Much research has been done since the IPCC's *Third Assessment Report*, but there has not been a similarly rigorous assessment of what is known with regard to temperature increase, precipitation, and weather-related events for the United States.[13] However, significant assessments will be completed in the near future. In particular, the IPCC is expected to release its *Fourth Assessment Report* throughout 2007.

While we were completing our review, the IPCC released a summary of the first of three components of its *Fourth Assessment Report*, which builds upon past IPCC assessments and incorporates new findings from the physical science research since the *Third Assessment Report*. The summary reports higher confidence in projected patterns of warming and other regional-scale features, including changes in wind patterns, precipitation, and some aspects of extreme events. In particular, the summary reports that it is very likely that hot extremes, heat waves, and heavy precipitation events will continue to become more frequent. Moreover, based on a range of models, IPCC's summary states that it is likely that future tropical cyclones (typhoons and hurricanes) will become more intense, with larger peak wind speeds and more heavy precipitation associated with ongoing increases in tropical sea surface temperatures. IPCC reports less confidence in projections of a global decrease in the number of tropical cyclones, and that the apparent increase in the proportion of very intense storms since 1970 in some regions is much larger than simulated by current models for that period. The full first component report was not publicly released prior to the issuance of our report and is expected some time after May 2007.

The other two components of the *Fourth Assessment Report* will cover impacts, adaptation, and vulnerability, and mitigation. These reports are expected to assess, among other things, key vulnerabilities and risks from climate change, including changes in extreme events. Additionally, the IPCC has committed to producing a capping report that is intended to synthesize and integrate material contained in the forthcoming reports, as well as other IPCC products.

In addition to the IPCC's work, CCSP is assessing potential changes in the frequency or intensity of weather-related events specific to North America in a report scheduled for release in 2008. According to a National Oceanic and Atmospheric Administration (NOAA) official and agency documents, the report will focus on weather extremes that have a significant societal impact, such as extreme cold or heat spells, tropical and extra-tropical storms, and droughts. Importantly, officials have said the report will provide an assessment of the observed changes in weather and climate extremes, as well as future projections.

More Frequent or More Severe Extreme Weather-Related Events Could Significantly Increase Insured Losses

Extreme weather-related events impact communities and economic activity by damaging homes and vehicles (e.g., see fig. 2), interrupting electrical service and business operations, or destroying crops. IPCC reported that the insurance industry—especially the property and casualty segment—are sensitive to the effects of weather-related events. This was highlighted in the Department of Commerce's comments on a draft of this report, which observed that altering either the frequency or severity of high impact extreme weather-related events could result in a significant increase in the risk posed to an insurer. For example, the agency said that what had been considered a 500-year event (i.e., its probability of occurring in a given year is 1 in 500) could shift under climate change to become a 100-year event (i.e., its probability of occurring in a given year is 1 in 100). Consequently, more frequent or more severe events have a greater potential for damage and, in turn, insured losses. As an official from Aon Re Australia, a large global reinsurer, reported, "The most obvious impact of climate change on the insurance sector will be the increase in insured property losses from extreme weather events."[14]

Source: FEMA.

Note: According to FEMA, the depth of the floodwaters underscores the extent of the damage caused by the 1993 Midwest flood. A total of 534 counties in nine states were declared for federal disaster aid.

Figure 2. July 1993 Flood Damage at Chesterfield Airport in St. Louis, Missouri.

Notably, the economic damages associated with some extreme weather-related events could increase at a greater rate in comparison with changes in the events themselves. Seemingly small changes in the characteristics of certain weather-related events can lead to substantial increases in damage. For example, recent work on hurricanes by researchers at the University of Colorado, the National Weather Service, and other institutions examined losses associated with hurricanes that made landfall in the United States since 1900.[15] Holding constant the increased population and development in coastal counties during this period, the study compared the economic damage of stronger storms with weaker storms, based on the Saffir-Simpson Hurricane

Scale.[16] The researchers found that stronger storms have caused many times more economic damages than weaker storms, as shown in figure 3. These findings are consistent with other independent analyses conducted by insurers and catastrophe modelers.

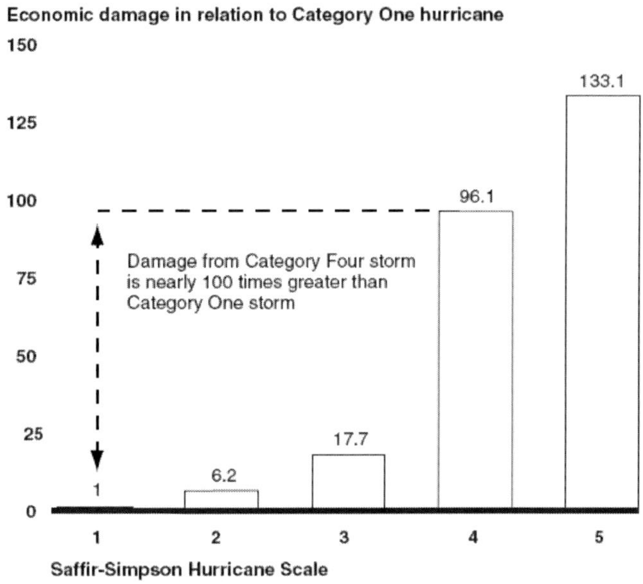

Figure 3. Economic Damages by Hurricane Category for U.S. Hurricanes Making Landfall, 1900-2005.

Moreover, public reports from several of the world's largest reinsurance companies and brokers underscore the potential for substantially increased losses. These reports note that, in addition to greater losses in absolute terms, the potential for greater variability in weather-related events could significantly enhance the volatility of losses.

Insured Weather-Related Losses Have Been Sizeable, and Federal Insurers' Exposure Has Grown Significantly

Taken together, insurers paid more than $320 billion in claims for weather-related losses between 1980 and 2005.[17] Claims varied significantly from year to year—largely due to the effects of catastrophic weather events such as hurricanes and droughts—but generally increased during this period. The growth in population in hazard-prone areas, and consequent real estate development and increasing real estate values, have generally increased insurers' exposure to weather-related events and help to explain their increased losses. Due to these and other factors, the federal insurance programs' liabilities have grown significantly, leaving the federal government increasingly vulnerable to the financial impacts of extreme events.

Claims Paid on Weather-Related Losses Totaled More Than $320 Billion between 1980 and 2005

Based on an examination of loss data from several different sources, insurers incurred more than $320 billion in weather-related losses from 1980 through 2005 (see fig. 4). Weather-related losses accounted for 88 percent of *all* property losses paid by insurers during this period. All other property losses, including those associated with earthquakes and terrorist events, accounted for the remainder. Weather-related losses varied significantly from year to year, ranging from just over $2 billion in 1987 to more than $75 billion in 2005.

Privately-Insured Losses

Of the $321.2 billion in weather-related loss payments we reviewed, private insurers paid $243.5 billion—over three-quarters of the total.[18] Figure 5 depicts the breakdown of these payments among key weather-related events. Of the $243.5 billion paid by private insurers, hurricanes accounted for $124.6 billion, or slightly more than half. Wind, tornados, and hail associated with severe thunderstorms accounted for $77 billion, or nearly one-third of the private total. Winter storms were associated with $25.1 billion, or about 10 percent.

Federally-Insured Losses

The two major federal insurance programs—NFIP and FCIC—paid the remaining $77.7 billion of the $321.2 billion in weather-related loss payments we reviewed.[19] Although the performance of both NFIP and FCIC is sensitive to weather, the two programs insure fundamentally different risks and operate in very different ways.

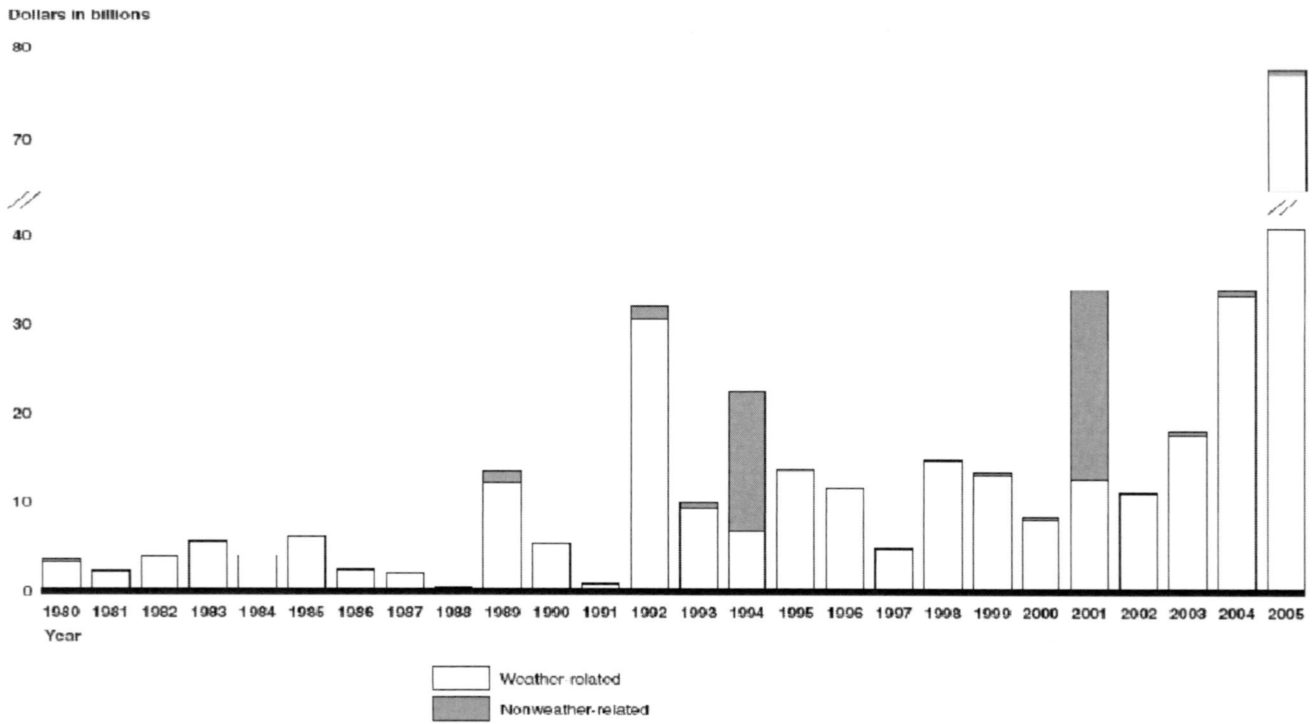

Sources: GAO analysis of PCS, NFIP, and FCIC data.

Figure 4. Annual Weather- and Nonweather-Related Insured Losses.

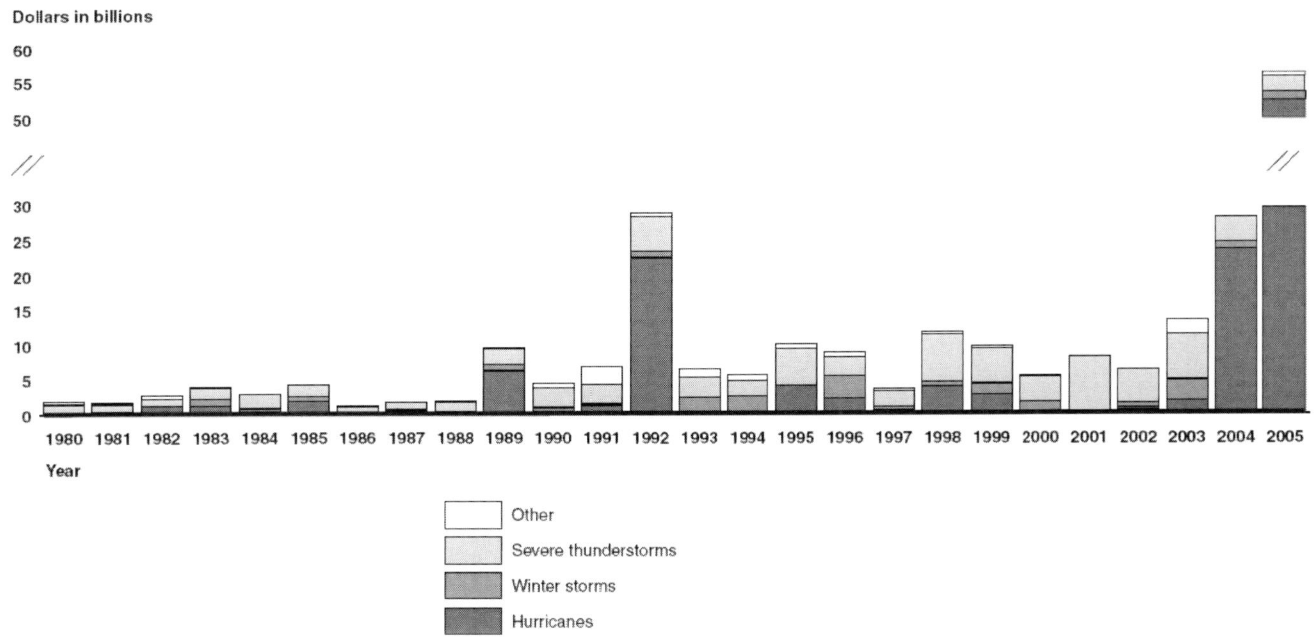

Source: GAO analysis of PCS data.

Figure 5. Weather-Related Losses Paid by Private Insurers.

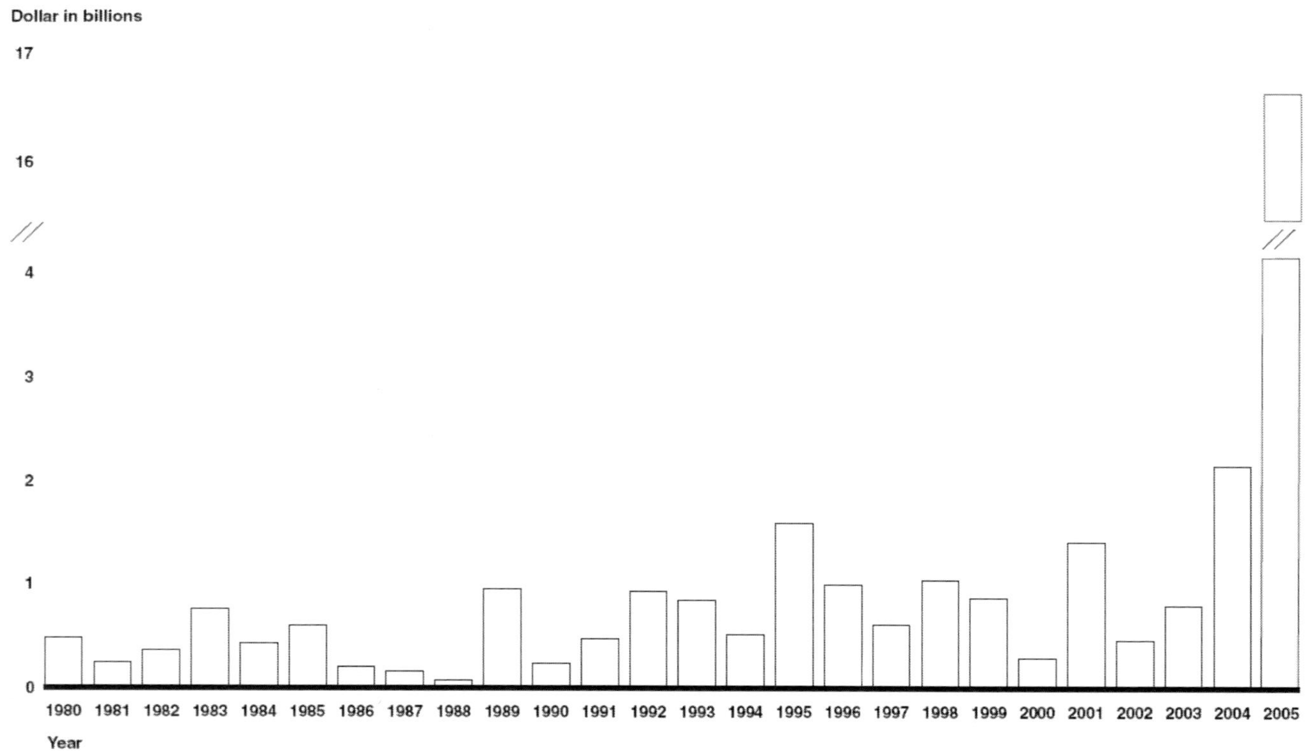

Source: GAO analysis of NFIP data.

Figure 6. Weather-Related Losses Paid by NFIP.

NFIP provides insurance for flood damage to homeowners and commercial property owners in more than 20,000 communities. Homeowners with mortgages from federally regulated lenders on property in communities identified as being in high flood risk areas are required to purchase flood insurance on their dwellings. Optional, lower cost flood insurance is also available under the NFIP for properties in areas of lower flood risk. NFIP offers coverage for both the property and its contents, which may be purchased separately.

NFIP claims totaled about $34.1 billion, or about 11 percent of all weather-related insurance claims during this period. As shown in figure 6, NFIP covers only one cause of loss—flooding. Claims averaged about $1.3 billion per year, but ranged from $75.7 million in 1988 to $16.7 billion in 2005.

FCIC insures commodities on a crop-by-crop and county-by-county basis based on farmer demand for coverage and the level of risk associated with the crop in a given region. Over 100 crops are covered by the program. Major crops, such as grains, are covered in almost every county where they are grown, and specialty crops, such as fruit, are covered only in some areas. Participating farmers can purchase different types of crop insurance, including yield and revenue insurance, and at different levels. For yield insurance, participating farmers select the percentage of yield of a covered crop to be insured and the percentage of the commodity price received as payment if the producer's losses exceed the selected threshold. Revenue insurance pays if actual revenue falls short of an assigned target level regardless of whether the shortfall was due to low yield or low commodity market prices.

Since 1980, FCIC claims totaled $43.6 billion, or about 14 percent of all weather-related claims during this period. FCIC losses averaged about $1.7 billion per year, ranging from $531.8 million in 1987 to $4.2 billion in 2002. Figure 7 shows the three causes of loss—drought, excess moisture, and hail—that accounted for more than three-quarters of crop insurance claims. In particular, drought accounted for $18.6 billion in losses, or more than 40 percent of all insured crop losses. Excess moisture totaled $11.2 billion, followed by hail with total claims of $4.2 billion. The remaining $9.6 billion in claims was spread among 27 different causes of loss, including frost and tornados.

Insured Losses Understate Total Economic Damage

Importantly, the insured loss totals used in our analysis do not account for all economic damage associated with weather-related events.[20] Specifically, data are not available for several categories of economic losses, including uninsured, underinsured, and self-insured losses. As we reported in 2005, FEMA estimates that one-half to two-thirds of structures in floodplains do not have flood insurance because the uninsured owners either are unaware that homeowners insurance does not cover flood damage, or they do not perceive a serious flood risk.[21] Furthermore, industry analysts estimate that 58 percent of homeowners in the United States are underinsured—that is, they carry a policy below the replacement value of their property—by an average of 21 percent.[22] Finally, some individuals and businesses have the means to "self-insure" their assets by assuming the full risk of any damage. Various public and private disaster relief organizations provide assistance to communities and individuals who suffer noninsured economic losses, although it was beyond the scope of this report to collect data on these losses. In particular, since 1989, $78.6 billion in federal disaster assistance funds have been obligated through the Disaster Relief Fund administered by FEMA, the largest—but not only—conduit for federal disaster assistance money provided in the wake of presidentially declared disasters and emergencies.

Overall, according to data obtained from Munich Re, one of the world's largest reinsurers, the type of insured losses we reviewed account for no more than about 40 percent of the total losses attributable to weather-related events.[23] NOAA's National Hurricane Center (NHC) uses a similar proportion to produce the agency's estimates of total economic damage attributable to hurricanes.[24] Although we did not independently evaluate the reliability of these estimates, subject area experts we spoke with confirmed that it was the best such estimate available and is widely used as an approximation of the relative distribution of losses. The difficulties we and others faced in accounting for weather-related losses were the subject of the National Academies' *The Impacts of Natural Disasters: A Framework for Loss Estimation*.[25] Reporting how best to account for the costs of natural disasters, including weather-related events, NAS found that there was no system in place in either the public or the private sectors to consistently capture information about the economic impact. Specifically, the NAS report found no widely accepted framework, formula, or method for estimating these losses. Moreover, NAS found no comprehensive clearinghouse for the disaster loss information that is currently collected. To that end, NAS recommended that the Office of Management and Budget, in consultation with FEMA and other federal agencies, develop annual, comprehensive estimates of the payouts for disaster losses made by federal agencies. Reviewing the status of this recommendation was beyond the scope of this report. Nevertheless, our experience with trying to obtain comprehensive information on disaster costs and losses underscores the NAS findings.

Catastrophic Weather-Related Events Help Explain the Significant Year-to-Year Variance in Losses

The largest insured losses in the data we reviewed were associated with catastrophic weather events. These events have a low probability of occurrence, but their consequences are severe. Notably, both crop insurers and other property insurers face the catastrophic risks posed by extreme events, although the nature of the events for each is very different. In the case of crop insurance, drought accounted for more than 40 percent of all insured losses from 1980 to 2005, and the years with the largest losses were associated with drought. Taken together, though, hurricanes were the most damaging event experienced by insurers in the data we reviewed. Although the United States experienced an average of only two hurricanes per year from 1980 through 2005, weather-related claims attributable to hurricanes totaled more than 45 percent of *all* weather-related insured losses—more than $146 billion. Moreover, these losses appear to be increasing.

Table 2. Insured Losses Associated with Hurricanes

Dollars in thousands			
	Categories One, Two	**Categories Three, Four, Five**	**Total**
1980s	$807,422 (11)	$9,905,042 (6)	**$10,712,464 (17)**
1990s	9,038,801 (11)	29,099,303 (8)	**38,138,104 (19)**
2000s	8,071,619 (7)	89,210,093 (7)	**97,281,712 (14)**
Total	**$17,917,842 (29)**	**$128,214,438 (21)**	**$146,132,280 (50)**

Sources: GAO analysis of PCS and NFIP data; NOAA (hurricane intensity classification).

Note: Totals do not include crop losses associated with hurricanes. Number of hurricanes associated with losses is included in parentheses. Hurricane classification was based on peak intensity at landfall.

In the data we reviewed, the years with the largest insured losses were generally associated with major hurricanes, defined as Category Three, Four, or Five on the Saffir-Simpson Hurricane Scale. Table 2 shows that, while 29 Category One and Two storms account for nearly $18 billion in losses, the 21 major storms account for over $126 billion in losses. In fact, claims associated with major hurricanes comprised 40 percent of all weather-related insured losses since 1980.

Importantly, hurricane severity is only one factor in determining the size of a particular loss—the location affected by the hurricane is also important. Generally, the more densely populated an area, the greater the extent of economic activity and accumulated value of the building stock. For instance, several studies have reviewed the economic impact of Hurricane Andrew, which tracked over Florida in 1992, in light of the dramatic real estate development that has occurred in the meantime. Researchers have normalized losses associated with the storm to account for societal changes by holding constant the value of building materials, real estate, and other factors so that the storm's impact could be adjusted to reflect contemporary conditions.[26] Hurricane Andrew, which resulted in roughly $25 billion in total economic losses in 1992, would have resulted in more than twice that amount—$55 billion—were it to have occurred in 2005, given current asset values.

Several recent studies have commented on the apparent increases in hurricane losses during this time period, and weather-related disaster losses generally, with markedly different interpretations. Some argue that loss trends are largely explained by changes in societal and economic factors, such as population density, cost of building materials, and the structure of insurance policies.[27] Others argue that increases in losses have been driven by changes in climate.[28]

To address this issue, Munich Re and the University of Colorado's Center for Science and Technology Policy Research jointly convened a workshop in Germany in May 2006 to assess factors leading to increasing weather-related loss trends.[29] The workshop brought together a diverse group of international experts in the fields of climatology and disaster research. Among other things, the workshop sought to determine whether the costs of weather-related events were increasing and what factors account for increasing costs in recent decades.

Workshop participants reached consensus on several points, including that analyses of long-term records of disaster losses indicate that societal change and economic development are the principal factors explaining observed increases in weather-related losses.[30] However, participants also agreed that changing patterns of extreme events are drivers for recent increases in losses and that additional increases in losses are likely given IPCC's projected increase in the frequency or severity of weather-related events.

Value at Risk in Federal Insurers' Portfolios Increased Significantly between 1980 and 2005

The growth in population in hazard-prone areas, and consequent real estate development and increasing real estate values, are leaving the nation increasingly exposed to higher insured losses. The close relationship between the value of the resource exposed to weather-related losses and the amount of damage incurred may have ominous implications for a nation experiencing rapid growth in some of its most disaster-prone areas. We reported in 2002 that the insurance industry faces potentially significant financial exposure due to natural catastrophes.[31] Heavily populated areas along the Northeast, Southeast, and Texas coasts have among the highest value of insured properties in the United States and face the highest likelihood of major hurricanes. According to insurance industry estimates, a large hurricane in Miami could cause up to $110 billion in insured losses with total losses as high as $225 billion. Several states—including Florida, California, and Texas—have established programs to help ensure that coverage is available in areas particularly prone to these events.[32]

AIR Worldwide, a leading catastrophe modeling firm, recently reported that insured losses should be expected to double roughly every 10 years because of increases in construction costs, increases in the number of structures, and changes in their characteristics. AIR's research estimates that, because of exposure growth, probable maximum catastrophe loss grew in constant dollars from $60 billion in 1995 to $110 billion in 2005, and it will likely grow to over $200 billion during the next 10 years.

Data obtained from both the NFIP and FCIC programs indicate the federal government has grown markedly more exposed to weather-related losses regardless of the cause. For example, NFIP data show that the number of policyholders and the value of the properties insured have both increased since 1980. Figure 8 shows the growth of NFIP's exposure in terms of both number of policies and the total coverage. The number of policies has more than doubled in this time period, from 1.9 million policies to more than 4.6 million. Moreover, although NFIP limits coverage to $250,000 for a personal structure and $100,000 for its contents, and $500,000 of coverage for a business structure and $500,000 on its contents, more policyholders' homes are approaching (or exceeding) these coverage limits. Accordingly, the total value covered by the program increased fourfold in constant dollars during this time from about $207 billion to $875 billion in 2005.

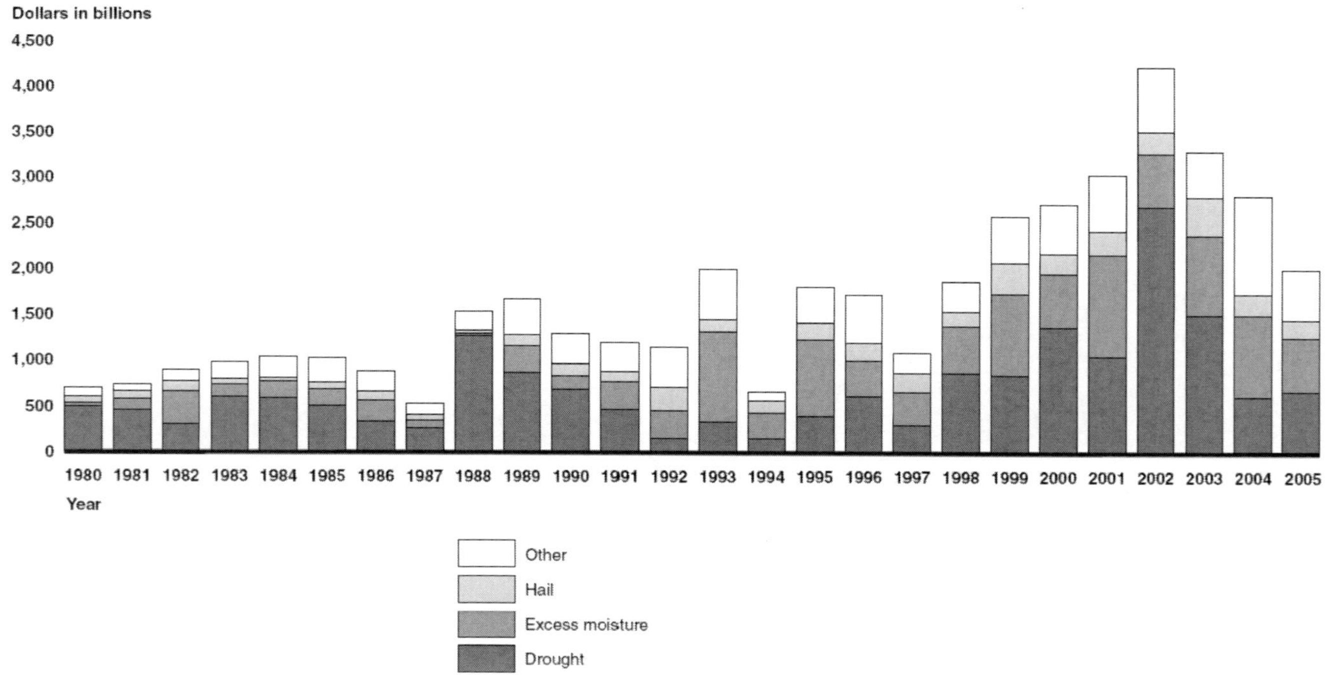

Source: GAO analysis of FCIC data.

Figure 7. Weather-Related Losses Paid by FCIC.

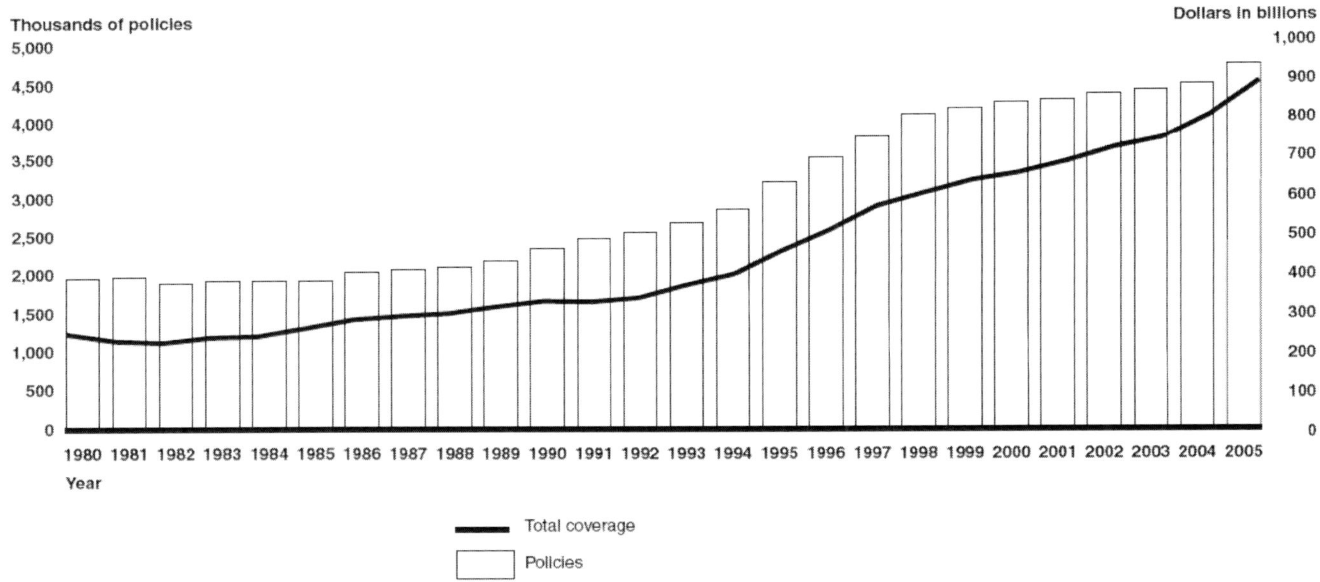

Source: GAO analysis of NFIP data.

Figure 8. NFIP Policies and Total Coverage.

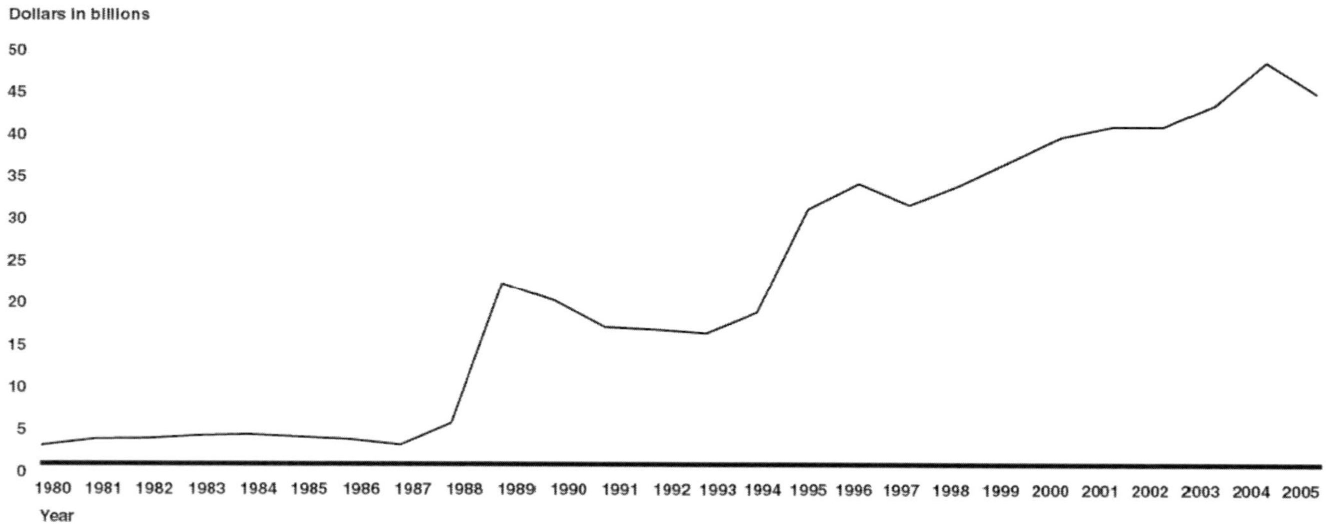

Source: GAO analysis of FCIC data.

Figure 9. FCIC Total Coverage.

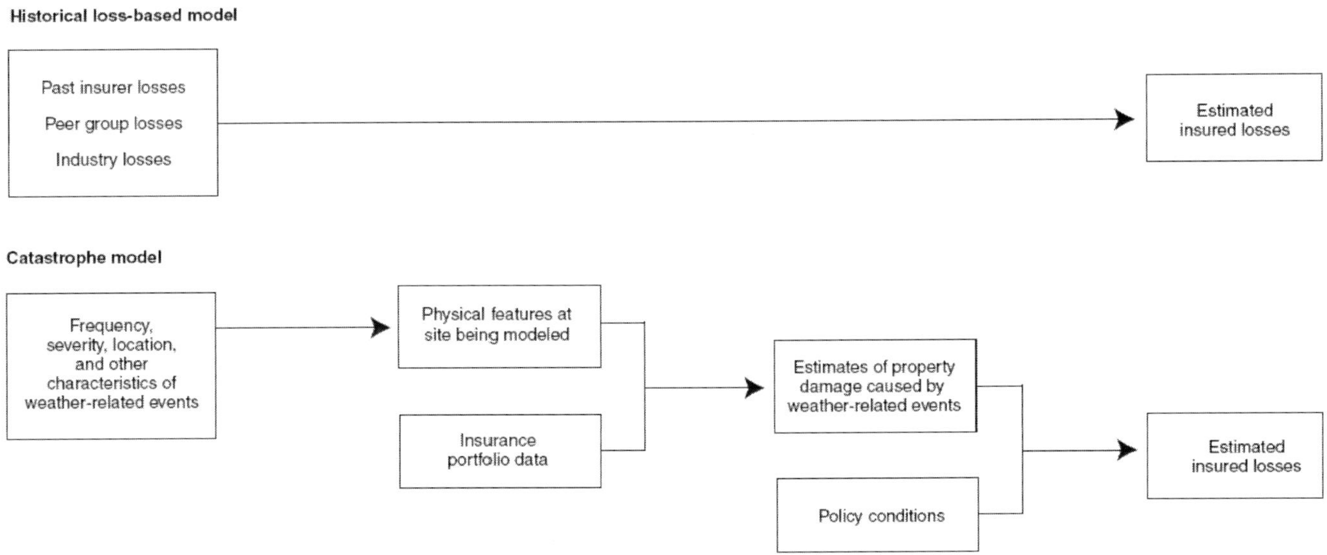

Sources: Adapted from the American Academy of Actuaries and Towers Perrin.

Figure 10. Modeling Potential Catastrophe Losses.

Similarly, RMA data show that FCIC has effectively increased its exposure base 26-fold during this period (in constant dollars). In particular, the program has significantly expanded the scope of crops covered and increased participation. Figure 9 shows the growth in FCIC exposure since 1980.[33]

A senior RMA official told us that the main implication of FCIC's growth is that the magnitude of potential claims, in absolute terms, is much greater today than in the past. For example, if the Midwest floods of 1993 were to occur today, losses would be five times greater than the $2 billion paid in 1993, according to RMA officials.

MAJOR PRIVATE AND PUBLIC INSURERS DIFFER IN HOW THEY MANAGE CATASTROPHIC RISKS ASSOCIATED WITH CLIMATE CHANGE

Although the relative contribution of event intensity versus societal factors in explaining the rising losses associated with weather-related events is still under investigation, both major private and federal insurers are exposed to increases in the frequency or severity of weather-related events associated with climate change. Nonetheless, major private and federal insurers are responding to this prospect differently.

Many large private insurers are incorporating some elements of near-term climate change into their risk management practices. Furthermore, some of the world's largest insurers have also taken a long-term strategic approach toward changes in climate.

On the other hand, for a variety of reasons, the federal insurance programs have done little to develop the kind of information needed to understand the programs' long-term exposure to climate change. We acknowledge the different mandate and operating environment in which the major federal insurance programs operate but believe that better information about the federal government's exposure to potential changes in weather-related risk would help the Congress identify and manage this emerging high-risk area; one which may not constitute an immediate crisis but which may pose an important longer term threat to the nation's welfare.

Major Private Insurers Prospectively Manage Potential Increases in Catastrophic Risk Associated with Climate Change

Extreme weather events pose a unique financial threat to private insurers' financial success because a single event can cause insolvency or a precipitous drop in earnings, liquidation of assets to meet cash needs, or a downgrade in the market ratings used to evaluate the soundness of companies in the industry. To prevent these disruptions, the American Academy of Actuaries (AAA)—the professional society that establishes, maintains, and enforces standards of qualification, practice, and conduct for actuaries in the United States—has outlined a five-step process for private insurers to follow to manage their catastrophic risk. These steps include the following:

- identifying catastrophic risk appetite by determining the maximum potential loss they are willing to accept;
- measuring catastrophic exposure by determining how vulnerable their total portfolio is to loss, both in absolute terms and relative to the company's risk management goals;
- pricing for catastrophic exposure by setting rates to collect sufficient premiums to cover their expected catastrophic loss and other expenses;
- controlling catastrophic exposure by reducing their policies in areas where they have too much exposure, or transferring risk using reinsurance or other mechanisms; and
- evaluating their ability to pay claims by determining the sufficiency of their financial resources to cover claims in the event of a catastrophe.

Additionally, insurers monitor their exposure to catastrophic weather-related risk using sophisticated computer models called "catastrophe models."[34] AAA emphasizes the shortcomings of estimating future catastrophic risk by extrapolating solely from historical losses and endorses catastrophe models as a more rigorous approach.[35] Catastrophe models incorporate the underlying trends and factors in weather phenomena and current demographic, financial, and scientific data to estimate losses associated with various weather-related events. According to an industry representative, catastrophe models assess a wider range of possible events than the historical loss record alone. These models simulate losses from thousands of potential catastrophic weather-related events that insurers use to better assess and control their exposure and inform pricing and capital management decisions. Figure 10 illustrates the difference between estimating future catastrophic losses using historical data versus catastrophe models.

To determine what major private insurers are doing to estimate and prepare for risks associated with potential changes in climate arising from natural or human factors, we contacted 11 of the largest private insurers operating in the U.S. property casualty insurance market. Representatives from each of the 11 major insurers we interviewed told us they use catastrophe models that incorporate a near-term higher frequency and intensity of hurricanes. Of the 11 private insurers, 6 specifically attributed the higher frequency and intensity of hurricanes to the Atlantic Multidecadal Oscillation, which—according to NOAA—is a 20- to 40-year climatic cycle of fluctuating temperatures in the north Atlantic Ocean. The remaining 5 insurers did not elaborate on the elements of climate change driving the differences in hurricane characteristics.

Industry reports indicate that insurance companies' perception of increased risk from hurricanes has prompted them to reduce their near-term catastrophic exposure, in both reinsurance and primary insurance coverage along the Gulf Coast and eastern seaboard. For example, a recent industry analysis from a leading insurance broker reported that reinsurance coverage is substantially limited in the southeastern United States and that reinsurance prices have more than doubled from 2005 to 2006, following a record-setting hurricane season.[36] According to the Insurance Information Institute, a leading source of information about the insurance industry, primary insurance companies have also raised prices in coastal states to cover rising reinsurance costs.[37] Additionally, a recent report co-authored by a major international insurance company cites several examples of large primary insurers either limiting coverage or withdrawing from vulnerable areas such as Florida,[38] the Gulf Coast, and Long Island.[39]

As private insurers limit their exposure, catastrophic risk is transferred to policyholders and the public sector. Insurance companies transfer risk to policyholders by increasing premiums and deductibles, or by setting lower coverage limits for policies. Insurers can also transfer risk to policyholders by passing along the mandatory participation costs of state-sponsored insurance plans.[40] For example, after the 2004 hurricane season, insurers assessed a surcharge of about 7 percent to every policyholder in Florida to recoup the cost of insurers' participation in the state-sponsored wind insurance plan. The public sector assumes management of weather-related risk at the local, state, and national level by providing disaster relief and recovery, developing mitigation projects, appropriating funds and, ultimately, providing insurance programs when private insurance markets are not sufficient or do not exist.

In addition to managing their aggregate exposure on a near-term basis, some of the world's largest insurers have also taken a long-term strategic approach to changes in catastrophic risk. For example, major insurance and reinsurance companies, such as Allianz, Swiss Re, Munich Re, and Lloyds of London, have published reports that advocate increased industry awareness of the potential risks of climate change and outline strategies to address the issue proactively. Moreover, 6 of the 11 private insurers we interviewed provided one or more additional activities they have undertaken when asked if their company addresses changes in climate through their weather-related risk management processes. These activities include monitoring scientific research (4 insurers), simulating the impact of a large loss event on their portfolios (3 insurers), and educating others in the industry about the risks of climate change (3 insurers), among others.

Furthermore, recent research on insurers' activities to address climate change outlines several other actions that private sector companies are taking, such as developing specialized policies and new products, evaluating risks to company stock investments, and disclosing to shareholders information about company-specific risks due to climate change.[41] Additionally, concern over the potential impacts of climate change on the availability and affordability of private insurance has led state insurance regulators to establish a task force to formally address the issue. The report, issued by the NAIC, is expected to be published in the summer of 2007.

Major Federal Insurers Have Taken Little Action to Prospectively Assess Potential Increases in Catastrophic Risk Associated with Climate Change

The goals of the major federal insurance programs are fundamentally different from those of private insurers. Specifically, whereas private insurers stress the financial success of their business operations, the statutes governing the NFIP and FCIC promote affordable coverage and broad participation by individuals at risk. Although both programs manage risk within their statutory guidelines, unlike the private sector, neither program is required to limit its catastrophic risk strictly within the programs' ability to pay claims on an annual basis. One important implication of the federal insurers' risk management approach is that they each have little reason to develop information on their long-term exposure to the potential risk of increased low-frequency, high-severity weather events associated with climate change.

The statutes governing the NFIP and FCIC promote broad participation over financial self-sufficiency in two ways: (1) by offering discounted or subsidized premiums to encourage participation and (2) by making additional funds available during high-loss years.[42] For example, discounted insurance premiums are available under the NFIP for some older homes situated within high flood risk areas where insurance would otherwise have been prohibitively expensive. FEMA is also authorized to borrow additional federal funds for the NFIP on an as-needed basis, subject to statutory limits, to cope with catastrophes.[43] One effect has been that the NFIP's exposure has expanded well beyond the ability to pay claims in high-loss years.

Similar to the discounted premiums offered by the NFIP, the FCIC's subsidized premiums are designed to make crop insurance available and affordable to as many participants as possible. For example, the FCIC is mandated to provide fully subsidized catastrophic coverage for producers in exchange for a minimal administrative fee, as well as partial subsidies for additional levels of coverage. Also like the NFIP, the FCIC is authorized to use additional federal funds on an as-needed basis during high-loss years—although, unlike the NFIP, the FCIC is not required to reimburse those additional funds.

Unlike the private sector, the NFIP and the FCIC can use additional federal funds, and so neither program is required to assess and limit its catastrophic risk strictly within its ability to pay claims on an annual basis. Instead, each program manages its risk to the extent possible, within the context of its broader purposes, in accordance with its authorizing statutes and implementing regulations.[44] For example, the FCIC uses coverage limits, exclusions, and premium rates to meet their statutory goal of a long-term loss ratio no greater than 1.075—including premium subsidies.[45] Although the program has experienced high-loss years that required additional federal funds, over time, these high-loss years have been offset by low-loss years, which have allowed the program to meet its goal and build reserves.[46]

By developing a goal to generate sufficient revenue to pay for an average loss year, the NFIP has also been able to generate a surplus in low-loss years despite borrowing funds in high-loss years. In the past, the program has been able to repay borrowed funds with interest to the Department of the Treasury, however, it is unlikely FEMA will be able to repay the nearly $21 billion borrowed following the 2005 hurricane season based on the program's current premium income.

Although neither program faces the potential of financial ruin like the private sector, both programs have occasionally attempted to estimate their aggregate losses from potential catastrophic events. For example, FCIC officials stated that they had modeled past events, such as the 1993 Midwest floods, using current participation levels to inform negotiations with private crop insurers over reinsurance terms. NFIP and FCIC officials explained that these efforts were informal exercises and were not performed on a regular basis. FCIC officials also said they use a hurricane model developed by NOAA to inform pricing decisions for some commodities such as citrus crops, according to FCIC officials. However, unlike the catastrophic risk faced by private insurers, hurricane damages have not been a primary source of crop insurance claims.

According to NFIP and FCIC officials, their risk management processes adapt to near-term changes in weather as they affect existing data. As one NFIP official explained, NFIP is designed to assess and insure against current—not future—risks. Over time, agency officials stated, this process has allowed their programs to operate as intended. However, unlike the private sector, neither program has conducted an analysis to assess the potential impacts of an increase in the frequency or severity of weather-related events on their program operations over the near- or long-term.

Information on Federal Agencies' Long-term Exposure to Catastrophic Risk Could Better Inform Congressional Decision Making

While comprehensive information on federal insurers' long-term exposure to catastrophic risk associated with climate change may not inform the NFIP's or FCIC's annual operations, it could nonetheless provide valuable information for the Congress and other policymakers who need to understand and prepare for fiscal challenges that extend well beyond the two programs' near-term operational horizons. We have highlighted the need for this kind of strategic information in recent reports that have expressed concern about the looming fiscal imbalances facing the nation. In one report, for example, we observed that, "Our policy process will be challenged to act with more foresight to take early action on problems that may not constitute an urgent crisis but pose important long-term threats to the nation's fiscal, economic, security, and societal future."[47] The prospect of increasing program exposure, coupled with expected increases in frequency and severity of weather events associated with climate change, would appear to pose such a problem.

Agency officials identified several challenges that could complicate their efforts to assess these impacts at the program level. Both NFIP and FCIC officials stated there was insufficient scientific information on projected impacts at the regional and local levels to accurately assess their impact on the flood and crop insurance programs. However, members of the insurance industry have analyzed and identified the potential risks climate change poses, despite similar challenges. Moreover, as previously discussed, both the IPCC and CCSP are expected to release significant assessments of the likely effect of increasing temperatures on weather events in coming months.

The experience of many private insurers, who must proactively respond to long-term changes in weather-related risk to remain solvent, suggests the kind of information that might be developed to help congressional and other policymakers in assessing current and alternative strategies. Specifically, to help ensure their future viability, a growing number of private insurers are actively incorporating the potential for climate change into their strategic level analyses. In particular, some private insurers have run a variety of simulation exercises to determine the potential business impact of an increase in the frequency and severity of weather events. For example, one insurer simulated the impact of large weather events occurring simultaneously. A similar analysis could provide the Congress with valuable information about the potential scale of losses facing the NFIP and FCIC in coming decades, particularly in light of the programs' expansion since 1980.

CONCLUSION

Recent assessments by leading scientific bodies provide sufficient cause for concern that climate change may have a broad range of long-term consequences for the United States and its citizens. While a number of key uncertainties regarding the timing, location, and magnitude of impacts remain, climate change has implications for the fiscal health of the federal government, which already faces other significant challenges in meeting its long-term fiscal obligations. NFIP and FCIC are two major federal programs which, as a consequence of both future climate change and substantial growth in exposure, may see their losses grow by many billions of dollars in coming decades.

We acknowledge that to carry out their primary missions, these public insurance programs must focus on the near-term goals of ensuring affordable coverage for individuals in hazard-prone areas. Nonetheless, we believe the two programs are uniquely positioned to provide strategic information on the potential impacts of climate change—information that would be of value to key decision makers charged with such a long-term focus. Most notably, in exercising its oversight responsibilities, the Congress could use such information to examine whether the current structure and incentives of the federal insurance programs adequately address the challenges posed by potential increases in the frequency and severity of catastrophic weather events. While the precise content of these analyses can be debated, the activities of many private insurers already suggest a number of strong possibilities that may be applicable to assessing the potential implications of climate change on the federal insurance programs.

RECOMMENDATION FOR EXECUTIVE ACTION

We recommend that the Secretary of Agriculture and the Secretary of Homeland Security direct the Administrator of the Risk Management Agency and the Under Secretary of Homeland Security for Emergency Preparedness to analyze the potential long-term implications of climate change for the Federal Crop Insurance Corporation and the National Flood Insurance Program, respectively, and report their findings to the Congress. This analysis should use forthcoming assessments from the Climate Change Science Program and the Intergovernmental Panel on Climate Change to establish sound estimates of expected future conditions. Key components of this analysis may include: (1) realistic scenarios of future losses under anticipated climatic conditions and expected exposure levels, including both potential budgetary implications and consequences for continued program operation and (2) potential mitigation options that each program might use to reduce their exposure to loss.

AGENCY COMMENTS AND OUR EVALUATION

We provided a draft of this report to the Departments of Agriculture (USDA), Commerce, Energy, and Homeland Security (DHS) for their review. DHS agreed via email with the report's recommendation, noting that conducting an assessment of the impact of climate change beyond FEMA's current statistical modeling (which is based on historical loss experience) could be helpful if resources were available to pursue such an analysis.

USDA also agreed with the report's recommendation, and commented on the presentation of several findings. In particular, USDA disagreed that it had thus far taken little action to prospectively assess potential increases in catastrophic risk associated with climate change. USDA explained that RMA does assess both the current and long-term exposure of the crop insurance program to catastrophic weather events, noting specifically that RMA (1) updates and publishes total program liability on a weekly basis and (2) estimates expected changes in liability up to 10 years ahead through its baseline projections. We acknowledge these activities, but believe it is important to note that they are limited in scope, focusing almost exclusively on retrospective measures of performance and not on the potential for increasingly frequent and intense weather-related events. These events, including drought and heavy precipitation events, are the key events acknowledged by USDA as posing catastrophic risk to the crop insurance program. Moreover, other RMA efforts to capture changes in weather-related risk rely on data reflecting what has been experienced in the past, not on what could be experienced in the future. The Department of Commerce neither agreed nor disagreed with the report's findings, but instead offered several comments on the presentation of several issues in the draft (particularly the depth in which several issues are discussed) as well as technical comments. We have incorporated these comments as appropriate. Notably, the Department of Commerce underscored the vulnerability of high-risk coastal development, stating that such vulnerabilities will only be amplified by climate change-related increases in the frequency or severity of weather-related events. Finally, the Department of Energy elected not to provide comments on the draft.

John B. Stephenson
Director, Natural Resources and Environment

APPENDIX I: SCOPE AND METHODOLOGY

We were asked us to (1) describe what is known about how climate change might affect insured and uninsured losses, (2) determine insured losses incurred by major federal agencies and private insurers and reinsurers resulting from weather-related events, and (3) determine what major federal agencies and private insurers and reinsurers are doing to assess and manage the potential risk of increased losses due to changes in the frequency and severity of weather-related events associated with climate change.

Scientific Literature

To address the first objective, we reviewed and summarized existing literature from significant policy-oriented scientific assessments from reputable international and national research organizations including the Intergovernmental Panel on Climate Change, National Academy of Sciences, and the multifederal agency U.S. Climate Change Science Program, as specified in table 3. It was beyond the scope of this report to independently evaluate the results of these studies.

Table 3. Key Policy-Oriented Scientific Assessments Reviewed by GAO

Organization	Publication
Intergovernmental Panel on Climate Change (IPCC)	• *Climate Change 2007: The Physical Science Basis, Summary for Policymakers (2007)* • *Climate Change 2001: Synthesis Report (2001)* • *Climate Change 2001: The Scientific Basis (2001)* • *Climate Change 2001: Impacts, Adaptation & Vulnerability* (2001)
Climate Change Science Program (CCSP)	• Temperature Trends in the Lower Atmosphere: Steps for Understanding and Reconciling Differences, Synthesis and Assessment Product 1.1 (2006)
National Academy of Sciences (NAS)	• *Surface Temperature Reconstructions for the Last 2,000 Years* (2006) • *Understanding and Responding to Climate Change: Highlights of National Academies Reports (2006)* • *Radiative Forcing of Climate Change: Expanding the Concept and Addressing Uncertainties (2005)* • *From Climate to Weather: Impacts on Society and Economy–Summary of a Forum, June 28, 2002, Washington, D.C. (2003)* • *Understanding Climate Change Feedbacks (2003)* • *Abrupt Climate Change: Inevitable Surprises (2002)* • *Climate Change Science: An Analysis of Some Key Questions* (2001)

Source: GAO.

Note: Publication year follows publication title in parentheses.

Insured Loss Data

To address the second objective, we analyzed insured loss data from January 1, 1980, through December 31, 2005, from the Federal Emergency Management Agency (FEMA) for the National Flood Insurance Program (NFIP); the Department of Agriculture's Risk Management Agency (RMA) for the Federal Crop Insurance Corporation (FCIC); and the Property Claim Services (PCS) for private property insurance. Through electronic testing and other means, we assessed the reliability of each of the data sets to determine whether the data were sufficiently reliable for our purposes. Specifically, we interviewed the sources for each of the data sets to gather information on how records were collected, processed, and maintained. Because not all catastrophes are weather-related, we excluded all events attributable to terrorist acts, tsunamis, earthquakes, and other nonweather-related losses, based on discussions with the data provider. To adjust for the general effects of inflation over time we used the chain-weighted gross domestic product price index to express dollar amounts in inflation-adjusted 2005 dollars. We reviewed any changes in data collection methodologies that have occurred over time, and evaluated the effect of any changes on our ability to report losses. We believe that these data are sufficiently reliable for the purpose of describing insured losses. We note, however, that these data likely understate the actual insured losses.

PCS

PCS data are estimates of insured losses, or claims paid by private insurance companies, for catastrophe loss events for the 50 states, as well as the District of Columbia, Puerto Rico, and the U.S. Virgin Islands. PCS defines "catastrophes" as events that, in their estimation, affect a significant number of policyholders and that cause more than $25 million in damages. To identify catastrophes, PCS reviews daily weather reports and wire service news stories to determine if potentially damaging weather has occurred anywhere in the nation. PCS contacts adjusters, insurance claims departments, or public officials to gather additional information about the scope of damage and potential insured losses for events. Damages associated with a single storm event are grouped together as a single catastrophe, even if they are separated by distance. PCS obtains its insured loss data from information reported by insurers. PCS estimates include losses under personal and commercial property insurance policies covering real property, contents, business interruption, vehicles, and boats. PCS estimates also typically include amounts paid by state wind pools, joint underwriting associations, and certain other residual market mechanisms, such as Fair Access to Insurance Requirements (FAIR) plans. However, PCS estimates do not include damage to uninsured or self-insured property including uninsured publicly owned property and utilities; losses involving agriculture, aircraft and property insured under NFIP or certain specialty lines (such as ocean marine), or loss adjustment expenses. Generally, PCS finalizes its estimates within 6 months of the occurrence of a PCS-identified catastrophe, according to company documents. PCS does not independently verify or audit the accuracy of the reported losses. Thus, loss totals are the best estimates of primary insurers compiled by PCS professionals, and may or may not accurately and completely reflect actual industry-insured losses. Nevertheless, PCS has determined their data to be very close to other independent estimates. PCS officials said that, when compared with state insurance commissioners' estimates based on all loss data from insurance companies following particularly large catastrophes, PCS data are within 3 to 5 percent of actual amounts. For the data used in our review, company officials told us that most estimates included in the data provided to us are final, except the 2005 hurricanes.

NFIP

NFIP data are actual claim payment totals, not estimated amounts. NFIP data represent the budget outlays that satisfy claims submitted by NFIP policyholders to their participating program companies. The companies report these data to the NFIP on a monthly basis. According to a senior program official, the Department of Homeland Security performs periodic audits of company records reported to NFIP. Although nearly all claims in the NFIP data we reviewed are considered closed by the agency (and, therefore, final), a small portion of claims associated with 2004 and 2005 hurricane season are not reflected in data we reviewed, according to the agency's database manager.

FCIC

The loss data provided by FCIC represent the actual amount paid to policyholders, not estimates. FCIC data represent the budget outlays that satisfy claims submitted by policyholders to their participating insurance companies. Participating insurance companies submit claims information for processing through a computerized validation system. Automated processing of claims information occurs annually for a period going back 5 years, but agency officials said that indemnities may have changed after automated processing closed in very specific cases, such as settlement of litigation or arbitration cases.

Identifying Insured Losses Associated with Hurricanes

To determine the insured losses associated with major and nonmajor hurricanes, we identified losses associated with hurricanes in both the PCS and NFIP data sets. We used the name and year of each hurricane to link loss records to information from the National Oceanic and Atmospheric Administration (NOAA) on the peak intensity of each hurricane at or near landfall.

Independent Studies

We supplemented our descriptive analysis with a review of existing literature and the views of subject area experts on the primary drivers of changes in the weather-related loss record in general. Given the data challenges faced by natural hazard researchers, the data sets used in these studies are generally different.

Interviews with Major Insurers

To address the third objective, we conducted semistructured interviews with officials from the NFIP, RMA, and a nonprobability sample of the largest private property/casualty primary insurance and reinsurance companies as defined by national market share. In the private sector, 11 out of 14 potential respondents elected to participate, drawing from companies in the United States, Europe, and Bermuda. Although the results from this sample should not be generalized to represent all insurance companies, the companies we interviewed represent about 45 percent of the total domestic insurance market. In developing our semistructured questionnaire, we reviewed existing literature on risk assessment and management practices, GAO guidance on risk management, and interviewed subject area experts knowledgeable about the insurance industry and federal insurance programs. Insurance industry experts included representatives from insurance brokers, catastrophe modeling firms, industry associations, the Insurance Information Institute, and academics. To reduce response error, we pretested our questions for clarity, relevancy, and sensitivity with representatives from several insurance industry associations, including the American Insurance Association, the National

Association of Mutual Insurance Companies, the Property Casualty Insurance Association of America, and the Reinsurance Association of America. On the basis of feedback from the pretests, we modified the questions as appropriate. We distinguished proactive risk management responses to climate change from other responses according to whether insurers indicated that they were adjusting their activities based on projected changes in underlying weather trends rather than adapting only as changes in weather conditions reveal themselves in historical data. During our interviews, some private insurers attributed their actions to changes in the Atlantic Multidecadal Oscillation (AMO).

Because NOAA considers the AMO to be a climatic cycle, we categorized the actions of these insurers as responding to climate change.

We asked the participating federal agencies and private insurance and reinsurance companies to identify individuals knowledgeable about their weather-related risk management practices for our interviews. Based on these criteria, we spoke with a range of senior officials and representatives that included actuaries, underwriters, catastrophe specialists, regulatory affairs and counsel. During the interviews, we asked a series of questions about risk assessment and management practices for weather-related risk, significant drivers of changes to past and future weather-related risk, respondents' perception of and actions to address climate change in their risk management processes, and risk management best practices that might be transferable to federal insurers.

We also interviewed officials from rating agencies, catastrophe modeling firms, insurance industry associations, the National Association of Insurance Commissioners, and universities to provide additional context for respondents' statements. To supplement our interviews, we reviewed documentary evidence of risk management practices from federal agencies, studies from subject area experts, industry reports, publicly available insurance company documents, and previous work from GAO to provide context and support for respondents' statements.

We performed our work between February 2006 and January 2007 in accordance with generally accepted government auditing standards.

APPENDIX II: NATIONAL FLOOD INSURANCE PROGRAM

Floods are the most common and destructive natural disaster in the United States. According to NFIP statistics, 90 percent of all natural disasters in the

United States involve flooding. Because of the catastrophic nature of flooding and the inability to adequately predict flood risks, private insurance companies largely have been unwilling to underwrite and bear the risk of flood insurance. As a result, flooding is generally excluded from homeowner policies that cover damages from other types of losses, such as wind, fire, and theft.

The NFIP was established in 1968 to address uninsured losses due to floods. Prior to the establishment of the NFIP, structural flood controls on rivers and shorelines (e.g., dams and levees) and disaster assistance for flood victims were the federal government's primary tools for addressing floods. The Mississippi River Commission, created in 1879 to oversee the development of a levee system to control the river's flow, was the first of these federal efforts to address flooding. Due to the limited effectiveness of structural flood controls, continued development in flood-prone areas, and a desire to reduce postdisaster assistance payments, the Congress began examining the feasibility of prefunding flood disaster costs via federal insurance in the 1950s. Although the first federal flood insurance program authorized by the Congress in 1956 failed due to lack of funding, a series of powerful hurricanes and heavy flooding on the Mississippi River in the early 1960s prompted the Congress to revisit the issue and direct the Department of Housing and Urban Development (HUD) to conduct a feasibility study of a federal flood insurance program. The 1966 HUD feasibility study helped lead to the passage of the National Flood Insurance Act of 1968,[1] which authorized the creation of the NFIP.[2]

Since its inception, the NFIP has undergone several major changes in response to significant flood events. Hurricane Agnes in 1972 led to the mandatory flood insurance requirements on certain persons in flood-prone areas included in the Flood Disaster Protection Act of 1973, which also significantly increased coverage limits in a further effort to increase participation.[3] Following the Midwest floods of 1993, the Congress enacted the National Flood Insurance Reform Act of 1994, which strengthened lender compliance requirements with mandatory purchase provisions requiring mortgage-holders in flood-prone areas to purchase flood insurance and prohibited flood disaster assistance for properties that had not maintained their mandatory coverage.[4] In 2004, recognizing that losses from repetitive flooding on some insured properties was straining the financial condition of the NFIP, the Congress passed the Flood Insurance Reform Act of 2004, which provided NFIP with additional tools to reduce the number and financial impact of these properties.[5] These tools include: increased authorization of funding for mitigation of repetitive loss properties and statutory authority to penalize policyholders who refuse government assistance to mitigate certain structures that have been substantially or repetitively damaged by flooding, among others. Recently, the Congress has begun exploring additional changes to the NFIP to address the financial and operational challenges presented by the 2005 hurricane season.

How the Program Works

FEMA, within the Department of Homeland Security (DHS), is responsible for the oversight and management of the NFIP.[6] Under this program, the federal government assumes the liability for covered losses and sets rates and coverage limitations, among other responsibilities.

The NFIP combines three elements: (1) property insurance for potential flood victims, (2) mapping to identify the boundaries of the areas at highest risk of flooding, and (3) incentives for communities to adopt and enforce floodplain management regulations and building standards (such as elevating structures) to reduce future flood damage. The effective integration of all three of these elements is needed for the NFIP to achieve its goals of

- providing property flood insurance coverage for a high proportion of property owners who would benefit from such coverage,
- reducing taxpayer-funded disaster assistance when flooding strikes, and
- reducing flood damage through floodplain management and the enforcement of building standards.

Over 20,000 communities across the United States and its territories participate in the NFIP by adopting and agreeing to enforce state and community floodplain management regulations to reduce future flood damage. In exchange, the NFIP makes federally backed flood insurance available to homeowners and other property owners in these communities. As of 2005, the program had over 4.9 million policyholders, representing about $875 billion in assets. Homeowners with mortgages from federally regulated lenders on property in communities identified to be in high flood risk areas are required to purchase flood insurance on their dwellings. Optional, lower cost coverage is also available under the NFIP to protect homes in areas of low to moderate risk. The mandated coverage protects homeowners' dwellings only; to insure furniture and other personal property items against flood damage, homeowners must purchase separate NFIP personal property coverage.

Prior to the 2005 hurricanes, NFIP had paid about $14.6 billion in flood insurance claims, primarily from policyholder premiums that otherwise would have been paid through taxpayer-funded disaster relief or borne by home and business owners themselves. According to FEMA, every $3 in flood insurance claims payments saves about $1 in disaster assistance payments, and the combination of floodplain management and mitigation efforts save about $1 billion in flood damage each year.

To make flood insurance available on "reasonable terms and conditions to persons who have need for such protection,"[7] the NFIP strikes a balance between the scope of the coverage provided and the premium amounts required to provide that coverage. Policy coverage limits arise from statute and regulation, including FEMA's standard flood insurance policy (SFIP), which is incorporated in regulation and issued to policyholders when they purchase flood insurance. As of 2006, FEMA estimated 26 percent of its policies were subsidized, and 74 percent were charged "full-risk premium" rates. In 1981, FEMA set the operating goal of generating premiums at least sufficient to cover losses and expenses relative to the "historical average loss year." However, the heavy losses from the 2005 hurricane season may increase the historical average loss year to a level beyond the expected long-term average. In light of this, FEMA is currently revisiting the use of the historical average loss year as a premium income target.

Risk Assessment Practices

The NFIP uses hydrologic models to estimate loss exposure in flood-prone areas, based on the method outlined in the 1966 HUD report, *Insurance and Other Programs for Financial Assistance to Flood Victims*.[8] These techniques of analysis were first developed by hydrologists and hydraulic engineers to determine the feasibility of flood protection.

The hydrologic method uses available data on the occurrence of floods and flood damages to establish both the frequency of flood recurrence and the damage associated with a flood of a given height. The NFIP augments available flood data with detailed engineering studies, simulations, and professional judgment to establish the scientific and actuarial basis for its risk assessment process and rates.

Flood-elevation frequency data for specific communities is published in Flood Rate Insurance Maps, which differentiate areas based on their flood risk. These maps are the basis for setting insurance rates, establishing floodplain management ordinances, and identifying properties where flood insurance is mandatory.

To estimate expected annual losses and determine the basis for rate setting, NFIP combines flood-elevation frequency data with depth-damage calculations to estimate a range of flood probabilities and associated damages. Each possible flood is multiplied by the expected damage should such a flood occur, and then each of these is added together. The total of each possible flood's damage provides an expected per annum percentage of the value of property damage due to flooding. This expected damage can then be converted to an expected loss per $100 of property value covered by insurance. This per annum expected loss provides the fundamental component of rate setting. Rates are also adjusted to incorporate additional expense factors, such as adjustment costs and deductibles.

Program Funding

To the extent possible within the context of its broader purposes, the NFIP is expected to pay operating expenses and flood insurance claims with premiums collected on flood insurance policies rather than with tax dollars. However, as we have reported, the program is not actuarially sound by design because the Congress authorized subsidized insurance rates to be made available for policies covering certain structures to encourage communities to join the program. As a result, the program does not collect sufficient premium income to build reserves to meet the long-term future expected flood losses.[9] FEMA has statutory authority to borrow funds from the Department of the Treasury to keep the NFIP solvent.[10] Prior to the 2005 hurricane season, FEMA had exercised its borrowing authority four times, when losses exceeded available fund balances. For example, FEMA borrowed $300 million to pay an estimated $1.8 billion on flood insurance claims resulting from the 2004 hurricane season. Following hurricanes Katrina, Rita, and Wilma, FEMA estimates it will need to borrow nearly $21 billion dollars to cover outstanding claims. Although FEMA has repaid borrowed funds with interest in the past, FEMA does not expect to be able to meet the $1 billion in annual interest payments for these borrowed funds.

APPENDIX III: FEDERAL CROP INSURANCE CORPORATION

In general, farm income is determined on the basis of farm production and prices, both of which are subject to wide fluctuations due to external factors. Because a substantial part of farming depends on weather, farm production levels can vary substantially on an annual basis. Commodity prices are also subject to significant swings due to supply and demand on the domestic and international markets. The Congress created FCIC in 1938 to administer a federal crop insurance program on an experimental basis to temper the weather effects of the dust bowl and the economic effects of the Great Depression.[1]

The federal crop insurance program protects participating farmers against financial losses caused by droughts, floods, or other natural disasters. Until 1980, the federal crop insurance program was limited to major crops in the nation's primary production areas. The Federal Crop Insurance Act of 1980 expanded crop insurance both in terms of crops and geographic areas covered.[2] The expansion was designed to allow the disaster assistance payment program provided by the government under previous farm bills to be phased out. To encourage participation, the 1980 act required a 30 percent premium subsidy for producers who purchased coverage up to the 65 percent yield level. Despite the subsidies, program participation remained low, and the Congress authorized several ad hoc disaster payments between 1988 and 1993. Congressional dissatisfaction with the size and frequency of these payments prompted the Congress to pass the Federal Crop Insurance Reform Act of 1994, which mandated participation in the crop insurance program as a prerequisite for other benefits, including agriculture price support payments.[3] The 1994 act also introduced catastrophic risk protection coverage, which compensated farmers for losses exceeding 50 percent of their average yield at 60 percent of the commodity price. Premiums for catastrophic risk protection coverage were completely subsidized, and subsidies for other coverage levels were also increased.

As part of the 1996 Farm Bill, the Congress created the Office of Risk Management under the U.S. Department of Agriculture (USDA), and USDA established RMA to administer the FCIC insurance programs, among other things.[4] The Congress also required the creation of a revenue insurance pilot project and repealed the mandatory participation provision of the 1994 Act. However, participation in the crop insurance program has not necessarily precluded the need for further disaster assistance. For example, due to low commodity prices in 1997 and multiple years of natural disasters, the Congress enacted an emergency farm financial assistance package totaling almost $6 billion in 1998, which included over $2 billion in crop disaster payments, and an $8.7 billion financial assistance package in 1999 that included $1.2 billion in crop disaster payments.

In 2000, the Congress enacted the Agricultural Risk Protection Act, which further increased subsidies for insurance above the catastrophic risk protection coverage level, subsidized a portion of the cost of revenue insurance products, improved coverage for farmers affected by multiple years of natural disasters, required pilot insurance programs for livestock farmers, and authorized pilot programs for growers of other commodities not currently covered, gave the private sector greater representation on the FCIC Board of Directors, reduced eligibility requirements for permanent disaster payment programs for noninsured farmers, and provided new tools for monitoring and controlling program abuses, among other provisions.[5] These changes required $8.2 billion in additional spending from fiscal years 2001 through 2005.

How the Program Works

RMA has overall responsibility for supervising the federal crop insurance program, which it administers in partnership with private insurance companies. Insurance policies are sold and completely serviced through approved private insurance companies that have their losses reinsured by USDA. These companies share a percentage of the risk of loss or opportunity for gain associated with each insurance policy written. In addition, RMA pays companies a percentage of the premium on policies sold to cover the administrative costs of selling and servicing these policies. In turn, insurance companies use this money to pay commissions to their agents who sell the policies and fees to adjusters when claims are filed. RMA oversees the development of new insurance products and the expansion of existing insurance products to new areas to help farmers reduce the chance of financial loss.

The USDA determines whether the federal crop insurance program will insure a commodity on a crop-by-crop and county-by-county basis, based on farmer demand for coverage and the level of risk associated with the crop in the region, among other factors. Over 100 crops are covered; major crops such as grains are covered in almost every county where they are grown, and specialty crops such as fruit are covered in some areas. For many commodities, producers may also purchase revenue insurance. Based on commodity market prices and the producer's production history, producers are assigned a target revenue level. The producer receives a payment if their actual revenue falls short of the target level, whether the shortfall was due to low yield or low prices. Premiums for revenue insurance are subsidized at the same level as traditional crop insurance policies.

Farmers' participation in the federal crop insurance program is voluntary, but the federal government encourages it by subsidizing the insurance premiums. Participating farmers are assigned a "normal" crop yield based on their past production history and a commodity price based on estimated market conditions. The producer selects both the percentage of yield to be covered and the percentage of the commodity price received as payment if the producer's losses exceed the selected threshold. Premium prices increase as levels of yield and price coverage rise. However, all eligible producers can receive fully subsidized catastrophic risk protection coverage that pays producers for losses exceeding 50 percent of normal yield, at a level equal to 55 percent of the estimated market price, in exchange for a $100 administrative fee. Producers who purchase this coverage can buy additional insurance at partially subsidized rates up to 85 percent of their yield and 100 percent of the estimated market price.

As an alternative, the Group Risk Plan provides coverage based on county yields rather than a producer's actual production history. If county yield falls below the producer's threshold yield (a percentage of the historical county yield), then the producer receives a payment.

Risk Assessment Practices

RMA's risk assessment/rate-setting methodology is complex because the risk of growing a particular crop varies by county, farm, and farmer. Because of all the possible combinations involved, hundreds of thousands of rates are in place. Each year, RMA follows a multistep process to establish rates for each crop included in the program. The process involves establishing base rates for each county crop combination and adjusting these basic rates for a number of factors, such as coverage and production levels. In addition, rates are adjusted to account for the legislated limitations in price increases.

For each crop, RMA extracts data on counties' crop experience from its historical database. The data elements for each crop, crop year, and county include (1) the dollar amount of the insurance coverage sold, (2) the dollar amount of the claims paid, and (3) the average coverage level. The historical data are adjusted to the 65 percent coverage level (the most commonly purchased level of coverage) so that liability and claims data at different coverage levels can be combined to develop rates. Using the adjusted data, FCIC computes the loss-cost ratio for each crop in each county. The loss-cost ratio is calculated by dividing the total claim payments by the total insurance in force; the result is stated as a percentage.[6] To reduce the impact a single year will have on the average loss-cost ratio of each county, RMA caps the adjusted average loss-cost ratio for any single year at 80 percent of all years.[7] To establish the base rate for each county, the average for all the years since 1975 is calculated using the capped loss-cost ratios and a weighting process to minimize the differences in rates among counties.

Rates are further adjusted by: a disaster reserve factor, a surcharge for catastrophic coverage for each crop based on pooled losses at the state level,[8] a prevented planting factor, farm divisions, crop type, and differences in both average yield and coverage levels.[9]

Program Funding

The crop insurance program is financed primarily through general fund appropriations and farmer-paid premiums. In addition to the premiums paid by producers, FCIC receives an annual appropriation to cover necessary costs for the program's premium subsidies, excess losses, delivery expenses, and other authorized expenses. According to USDA budget documents, for fiscal year 2005, insurance premium and administrative fee revenue from farmers was approximately $2.1 billion, and gross claims equaled almost $3.3 billion. Total government operating costs in fiscal year 2005 were approximately $3 billion.

RMA is required to set crop insurance premiums at actuarially sufficient rates, defined as a long-run loss ratio target of no more than 1.075. From its initial expansion in 1981 through 1994, the crop insurance program had an average loss ratio of 1.47 and paid roughly $3.2 billion in claims excess of subsidized premium income during that period.[10] From 1995 to 2005, the program had an average loss ratio of 0.91, and collected roughly $2.7 billion in subsidized premium excess of claims during that period. Excluding subsidies and measuring performance on the basis of a producer premium, from 1981 to 1994, the crop insurance program averaged a loss ratio of 1.93 and paid roughly $5.2 billion in claims excess of producer premium over that period;

from 1995 to 2005, the program averaged a loss ratio of 2.15 and paid roughly $14.2 billion in claims excess of a producer premium during that period.

Generally, producers can purchase crop insurance to insure up to 85 percent of their normal harvest (yield), based on production history. In 2007, the USDA expects the FCIC to provide $48 billion in risk protection on 287 million acres nationwide, which represents approximately 80 percent of the nation's acres planted to principal crops. The USDA estimates this level of coverage will cost the federal government $4.2 billion in 2007.

APPENDIX IV: CONSENSUS STATEMENT AMONG PARTICIPANTS AT 2006 MUNICH RE WORKSHOP

Munich Re, one of the world's largest reinsurance companies, and the University of Colorado jointly convened an international workshop on climate change and disaster loss trends in May 2006 in Hohenkammer, Germany. The workshop brought together 32 experts in the fields of climatology and disaster research from 13 countries. White papers were prepared and circulated by 25 participants in advance of the workshop and formed the basis of the discussions. In the course of the event, participants developed a list of statements that each represent a consensus among participants on issues of research and policy as related to the workshop's two central organizing questions: (1) What factors account for increasing costs of weather related disasters in recent decades? and (2) What are the implications of these understandings, for both research and policy?

Consensus (unanimous) statements of the workshop participants:

1) Climate change is real, and has a significant human component related to greenhouse gases.
2) Direct economic losses of global disasters have increased in recent decades with particularly large increases since the 1980s.
3) The increases in disaster losses primarily result from weather related events, in particular storms and floods.
4) Climate change and variability are factors which influence trends in disasters.
5) Although there are peer reviewed papers indicating trends in storms and floods there is still scientific debate over the attribution to

anthropogenic climate change or natural climate variability. There is also concern over geophysical data quality.

6) IPCC (2001) did not achieve detection and attribution of trends in extreme events at the global level.
7) High quality long-term disaster loss records exist, some of which are suitable for research purposes, such as to identify the effects of climate and/or climate change on the loss records.
8) Analyses of long-term records of disaster losses indicate that societal change and economic development are the principal factors responsible for the documented increasing losses to date.
9) The vulnerability of communities to natural disasters is determined by their economic development and other social characteristics.
10) There is evidence that changing patterns of extreme events are drivers for recent increases in global losses.
11) Because of issues related to data quality, the stochastic nature of extreme event impacts, length of time series, and various societal factors present in the disaster loss record, it is still not possible to determine the portion of the increase in damages that might be attributed to climate change due to greenhouse gas emissions.
12) For future decades the IPCC (2001) expects increases in the occurrence and/or intensity of some extreme events as a result of anthropogenic climate change. Such increases will further increase losses in the absence of disaster reduction measures.
13) In the near future the quantitative link (attribution) of trends in storm and flood losses to climate changes related to greenhouse gas emissions is unlikely to be answered unequivocally.
14) Adaptation to extreme weather events should play a central role in reducing societal vulnerabilities to climate and climate change.
15) Mitigation of greenhouse gas emissions should also play a central role in response to anthropogenic climate change, though it does not have an effect for several decades on the hazard risk.
16) We recommend further research on different combinations of adaptation and mitigation policies.
17) We recommend the creation of an open-source disaster database according to agreed-upon standards.
18) In addition to fundamental research on climate, research priorities should consider needs of decision makers in areas related to both adaptation and mitigation.

19) For improved understanding of loss trends, there is a need to continue to collect and improve long-term and homogenous data sets related to both climate parameters and disaster losses.
20) The community needs to agree upon peer reviewed procedures for normalizing economic loss data.

End Notes

[1] Insurers use the term "loss" to refer to the dollar value of approved or settled claims arising from damages incurred by a policyholder. For the purposes of this report, weather-related loss refers to the dollar value of claims made on damage attributable to weather-related events. "Loss" does not account for premium or other income, deductibles, co-payments, or damages in excess of coverage.

[2] More specifically, we used the Intergovernmental Panel on Climate Change definition, which refers to climate change as a statistically significant variation in either the mean state of the climate or in its variability, persisting for an extended period (typically decades or longer). Climate change may be due to natural factors (e.g., internal processes or external forcings such as solar variations or heavy volcanic activity), or to persistent human-induced changes in the composition of the atmosphere or land use patterns.

[3] The National Association of Insurance Commissioners is an organization of insurance regulators from the 50 states, the District of Columbia, and the five U.S. territories.

[4] Federal insurance programs are not designed to earn financial profits.

[5] To insure a risk, private insurers must be able to both estimate an event's occurrence and its associated damages and be able to set premiums sufficient to cover their risk and earn a profit. In some cases, insurers may be prevented from charging sufficient premiums due to state regulatory actions.

[6] See appendixes II and III for additional information on how these programs operate, how they assess risk, and how they are funded.

[7] Appendix I contains additional information on the specific assessments we reviewed. CCSP is a multiagency effort to coordinate federal climate change science that is responsible for preparing a series of 21 climate science synthesis and assessment products (SAP) for the United States by 2008.

[8] This estimate comes from a recently released summary of a key component of IPCC's *Fourth Assessment Report* of the state of climate science, which reported an updated 100- year linear trend (1906-2005) of 0.74 degrees Celsius—larger than the corresponding 0.6 degrees Celsius reported in the 2001 *Third Assessment Report*.

[9] For the purposes of this report, extreme weather-related events are those with a low frequency of occurrence, but that cause severe damage, such as hurricanes, drought, winter storms, tornadoes, wildfires, and floods, among others.

[10] IPCC narrowed its range of projected warming in its recently released summary from the corresponding range of 1.4 to 5.8 degrees Celsius reported in the 2001 *Third Assessment Report*. Although these two sets of projections are broadly consistent, they are not directly comparable. IPCC notes in the summary that the new range is more advanced in that it provides best estimates and an assessed likelihood range. It also relies on a larger number of climate models of increasing complexity and realism, as well as new information regarding

the nature of feedbacks from the carbon cycle and constraints on climate response from observations.

[11] National Research Council, *Surface Temperature Reconstructions for the Last 2,000 Years* (Washington, D.C.: 2006).

[12] Likelihoods for projected changes are defined by the following conditions set by the IPCC: "very likely" indicates that a number of models have been analyzed for such a change, all those analyzed show it in most regions, and it is physically plausible; and "likely" indicates that theoretical studies and those models analyzed show such a change, but only a few models are configured in such a way as to reasonably represent such changes.

[13] The most recent national assessment for the United States, entitled *Climate Change Impacts on the United States*, was forwarded by a federal advisory committee to the Congress and the President in 2000 as required by the Global Change Research Act of 1990. We reported in 2005 that the subsequent assessment was not submitted in November 2004 as required by the act. Instead, according to the Department of Commerce, CCSP has committed to issuing 21 shorter reports by 2008. See GAO, *Climate Change Assessment: Administration Did Not Meet Reporting Deadline*, GAO-05-338R (Washington, D.C.: Apr. 14, 2005).

[14] Andrew Dlugolecki, *The Changing Risk Landscape: Implications for Insurance Risk Management* (1999) http://www.aon.com.au/pdf/reinsurance/Aon_Climate_Change.pdf (downloaded Jan. 8, 2007).

[15] See Roger Pielke, Jr., et al., *Normalized Hurricane Damages in the United States: 1900- 2005* (2007), accessed via http://sciencepolicy.colorado.edu/publications/special/normalized_hurricane_damages.html (downloaded Jan. 8, 2007).

[16] The Saffir-Simpson hurricane intensity category system was developed in the 1970s to calculate the destructive force of hurricanes. The scale ranges from Category One to Category Five, with Category Five being the most severe. For example, Category Three hurricanes have winds of 111 to 130 mph, whereas Category Five hurricanes have winds greater than 155 mph.

[17] Data throughout this section are presented in constant 2005 dollars to allow for a comparison of the dollar value of losses over time and are not otherwise adjusted. See appendix I for more information on data used in this report.

[18] Property Claim Services (PCS), an authority on insured property losses, maintains a database of estimated losses determined to be "catastrophes"—that is, loss events larger than $25 million that affect a significant number of policyholders. PCS estimates include losses under personal and commercial property insurance policies and typically include payments made on behalf of state-administered risk pools. PCS data are described in greater detail in appendix I.

[19] Appendixes II and III provide additional information about the structure and operation of FCIC and NFIP. Importantly, totals only reflect what was paid during this time—some losses incurred in 2005 may be omitted from this data set.

[20] Weather-related damages are also responsible for many indirect and non-market impacts that are not entirely accounted for, if at all, in economic terms, such as environmental damage. See NAS, *The Impacts of Natural Disasters: A Framework for Loss Estimation* (Washington, D.C.: 1999), 55-64.

[21] GAO, *Catastrophe Risk: U.S. and European Approaches to Insure Natural Catastrophe and Terrorism Risks*, GAO-05-199 (Washington, D.C.: Feb. 28, 2005), 61.

[22] Estimate was produced by Marshall & Swift/Boeckh, a leading supplier of local building cost information, residential and commercial property valuation services for the property and

casualty insurance sector in the United States. GAO did not independently evaluate the reliability of this estimate.

[23] Munich Re, *Topics 2000: Natural Catastrophes—the Current Position*. Geoscience Research Group (Munich, Germany: 1999).

[24] NHC estimates total losses by extrapolating from insured losses by assuming they account for approximately 50 percent of total losses.

[25] NAS (1999), 1.

[26] A normalization provides an estimate of the damage that would occur if storms from the past affected the same location under the societal conditions of another year.

[27] See, for example, Roger A. Pielke, Jr., "Disasters, Death, and Destruction: Making Sense of Recent Calamities," *Oceanography*, vol. 19, no. 2 (2006); Stanley A. Changnon et al., "Human Factors Explain the Increased Losses from Weather and Climate Extremes," *Bulletin of the American Meteorological Society*, vol. 81, no. 3 (2000); and Roger A. Pielke, Jr., and Christopher W. Landsea, "Normalized Hurricane Damages in the United States: 1925–95," *Weather and Forecasting*, vol. 13 (1998).

[28] See, for example, Evan Mills, Richard J. Roth, Jr., and Eugene Lecomte, *Availability and Affordability of Insurance Under Climate Change: A Growing Challenge for the U.S.* (Boston, Mass.: December 2005); Paul Epstein and Evan Mills, eds., *Climate Change Futures: Health, Ecological, and Economic Dimensions* (Boston, Mass.: November 2005); and Cynthia Rosenzweig et al., "Increased Crop Damage in the U.S. from Excess Precipitation Under Climate Change," *Global Environmental Change*, vol. 12 (2002).

[29] Peter Höppe and Roger Pielke, Jr., eds., *Report of the Workshop on Climate Change and Disaster Losses: Understanding and Attributing Trends and Projections*, Hohenkammer, Germany, May 25-26, 2006 (Munich, Germany: October 2006).

[30] Consensus statements agreed to at the workshop are listed in their entirety in appendix IV.

[31] GAO, *Catastrophe Insurance Risks: The Role of Risk-Linked Securities and Factors Affecting Their Use*, GAO-02-941 (Washington, D.C.: Sept. 24, 2002), 3.

[32] Past GAO work provided information on the Florida Hurricane Catastrophe Fund, California Earthquake Authority, and the Texas Windstorm Insurance Association. See GAO-02-941 and GAO, *Catastrophe Insurance Risks: Status of Efforts to Securitize Natural Catastrophe and Terrorism Risk*, GAO-03-1033 (Washington, D.C.: Sept. 24, 2003).

[33] To maintain comparability with other data, GAO did not adjust these data for changes in agricultural prices.

[34] There are three main catastrophe modeling firms: AIR Worldwide, Risk Management Solutions, and EQECAT. Although many of the insurers we interviewed use models from these firms, two of the eleven insurers have developed their own catastrophe models.

[35] American Academy of Actuaries, *Catastrophe Exposures and Insurance Industry Catastrophe Management Practices* (Washington, D.C.: American Academy of Actuaries, June 10, 2001), http://www.actuary.org/pdf/casualty/catastrophe_061001.pdf (downloaded Jan. 3, 2007), 10-12.

[36] Guy Carpenter, *The World Catastrophe Reinsurance Market: Steep Peaks Overshadow Plateaus* (New York, N.Y.: Guy Carpenter, September 2006), http://www.guycarp.com /portal/extranet/insights/reports.html?vid=30 (downloaded Jan. 3, 2007).

[37] Insurance Information Institute, *Catastrophes: Insurance Issues* (New York, N.Y.: Insurance Information Institute, November 2006), http://www.iii.org/media (downloaded Jan. 3, 2007).

[38] Allianz Group and World Wildlife Fund, *Climate Change and Insurance: An Agenda for Action in the United States* (New York, N.Y.: Allianz Group and World Wildlife Fund,

October 2006), http://www.allianz.com/en/allianz_group/sustainability/insight/studies_and_reports/page1. html?hits=reports (downloaded Jan. 4, 2007).

[39] The report notes that these decisions were due, in part, to state restrictions on rate increases that are designed to maintain insurance prices that are affordable, but may not accurately reflect the true potential for loss faced by the insured.

[40] Thirty-one states have FAIR plans, and six southern states have state-sponsored wind insurance plans that pool resources from insurers to cover the cost of coverage for their participants.

[41] Evan Mills and Eugene Lecomte, *From Risk to Opportunity: How Insurers Can Proactively and Profitably Manage Climate Change* (Boston, MA: Ceres, August 2006), http://www.ceres.org/pub/docs/Ceres_Insurance_Climate_%20Report_082206.pdf (downloaded Jan. 3, 2007), 34.

[42] Note that the federal government covers most, but not all, payments in the event of loss under the FCIC—insurance providers also share in the risk, as described in detail in appendix III.

[43] The Congress increased the NFIP's borrowing authority from $1.5 billion to approximately $20.8 billion in the wake of unprecedented losses associated with the 2005 hurricane season.

[44] A detailed description of each program's risk management practices can be found in appendixes II and III for the NFIP and FCIC, respectively.

[45] Loss ratio, an indicator used to evaluate program performance, is calculated by dividing claims paid by total premiums collected. A loss ratio greater than 1.00 indicates that the program paid more in claims than was collected in premiums

[46] The FCIC's average loss ratio from 1995 through 2005 was 0.91. From 1981 through 1994, it was 1.47. See appendix III for more information on the FCIC's performance.

[47] GAO, *21st Century Challenges: Reexamining the Base of the Federal Government*, GAO-05-325SP (Washington, D.C.: February 2005), 77.

End Notes for Appendix II

[1] Pub. L. No. 90-448, 82 Stat. 573.

[2] Senate Committee on Banking and Currency, *Insurance and Other Programs for Financial Assistance to Flood Victims*, 89th Cong., 2d Sess., 1966, Committee Print.

[3] Pub. L. No. 93-234, 87 Stat. 975 (1973).

[4] Pub. L. No. 103-325, 108 Stat. 2255 (1994).

[5] The Bunning-Bereuter-Blumenauer Flood Insurance Reform Act of 2004, Pub. L. No. 108- 264, 118 Stat. 712.

[6] In March 2003, FEMA and its approximately 2,500 staff became part of the Department of Homeland Security (DHS). Most of FEMA—including its Mitigation Division, which is responsible for administering the NFIP—is now part of the department's Emergency Preparedness and Response Directorate. However, FEMA retained its name and individual identity within the department. Under a reorganization plan proposed by the current Secretary of DHS, the Emergency Preparedness and Response Directorate would be abolished, and FEMA would report directly to the Undersecretary and Secretary of DHS.

[7] 42 U.S.C. § 4001(a)(4).

[8] Senate Committee on Banking and Currency, *Insurance and Other Programs for Financial Assistance to Flood Victims*.

[9] GAO, *Flood Insurance: Information on the Financial Condition of the National Flood Insurance Program*, GAO-01-992T (Washington, D.C.: July 19, 2001).

[10] See 42 U.S.C. § 4016.

End Notes for Appendix III

[1] Federal Crop Insurance Act, tit. V, 52 Stat. 72 (1938) (codified as amended at 7 U.S.C. §§ 1501-1524).

[2] Pub. L. No. 96-365, 94 Stat. 1312 (1980).

[3] Pub. L. No. 103-354, 108 Stat. 3178 (1994).

[4] Pub. L. No. 104-127, 110 Stat. 888 (1996).

[5] Pub. L. No. 106-224, 114 Stat. 358 (2000).

[6] For example, if the claims paid in 1 year totaled $7.36 and the insurance in force was $100, the loss-cost ratio is 7.36 percent. The percentage represents the rate that would need to be charged per $100 of insurance coverage if total premiums are to equal the total claim payments for that year. In this example, the 7.36 percent indicates that a rate of $7.36 was required per $100 of insurance coverage sold.

[7] The excess of losses above the capped amount is pooled at the state level and reallocated to the counties. According to FCIC, this procedure is intended to reduce the variation of rates from one year to the next.

[8] The surcharge is established by pooling the amount of insurance in force and the claim payments for capped years with the highest loss-cost ratios in each county that were not factored into the county unloaded rates at the state level. These data are used to calculate a statewide surcharge for catastrophic coverage (pooled claims payments divided by pooled insurance in force). If the pooled losses at the state level exceed five points, the excess is returned to the counties and included in the county unloaded rate.

[9] Prevented planting factor adds a provision for losses due to crops that were never planted because of external factors not directly related to yield loss.

[10] The Federal Crop Insurance Reform Act of 1994 mandated participation in the program to receive other commodity support payments, although this requirement was rescinded in 1996.

INDEX

A

abuse, 4
accountability, 10
accounting, 64
adaptation, 10, 16, 18, 26, 56, 95
adjustment, 83, 89
adverse event, 9
age, 84
agencies, 2, 5, 10, 19, 20, 22, 27, 29, 33, 42, 45, 46, 64, 80, 85
agency actions, 18
Agricultural Research Service(ARS), 3, 18
agriculture, 3, 13, 16, 18, 22, 27, 28, 29, 32, 53, 83, 90
annual rate, 17
appetite, 73
appropriations, 5, 11, 19, 34, 35, 93
arbitration, 84
ARS, 3, 18
assessment, 14, 30, 42, 44, 47, 51, 53, 54, 55, 56, 79, 96, 97
assets, 64, 73, 87
atmosphere, 53, 96
attribution, 94, 95
audit(s), 6, 33, 46, 83, 85
authority, 20, 54, 89, 97, 99
awareness, 75

B

barriers, 10, 26
base, 8, 19, 21, 30, 37, 72, 92
base rate(s), 92, 93
base year, 19
benchmarks, 38
benefits, 23, 90
bonds, 19
breakdown, 59
building code, 26, 39
businesses, 4, 13, 17, 39, 45, 49, 64
buyer, 37

C

carbon, 53, 97
carbon dioxide, 53
cash, 38, 73
catastrophe bonds, 20, 37
catastrophes, 67, 76, 82, 97
Census, 14
challenges, vii, 1, 2, 3, 5, 9, 16, 20, 23, 24, 25, 27, 29, 30, 38, 49, 77, 78, 79, 84, 87
clarity, 84
classes, 35
classification, 65
climate change, vii, viii, 1, 2, 3, 4, 5, 6, 7, 9, 10, 14, 15, 16, 17, 18, 19, 20, 21, 22, 23, 24, 25, 26, 27, 28, 29, 30, 31, 32, 33, 36,

37, 38, 39, 41, 42, 45, 46, 48, 51, 56, 72, 74, 75, 76, 77, 78, 79, 80, 85, 94, 95, 96
climate change issues, 32
climate extremes, vii, 1, 56
coastal communities, 7
combustion, 53
commercial, 63, 82, 97
commodity, 63, 90, 91, 92, 100
communities, 3, 8, 10, 18, 26, 27, 30, 35, 39, 44, 50, 56, 63, 64, 87, 89, 95
community, 8, 22, 30, 35, 39, 53, 87, 96
complex interactions, 53
complexity, 54, 96
compliance, 86
composition, 96
computer, 24, 47, 48, 51, 53, 54, 73
conference, 31
Congress, 5, 42, 43, 48, 50, 72, 77, 78, 79, 86, 89, 90, 91, 97, 99
Congressional Budget Office, 21
consensus, 66, 94, 98
conservation, 18, 27
construction, 26, 27, 28, 35, 67
consumers, 7
cost, vii, 5, 19, 21, 29, 37, 38, 41, 44, 63, 66, 74, 87, 91, 93, 94, 97, 99, 100
counsel, 85
covering, 3, 8, 15, 20, 22, 27, 36, 82, 89
crop(s), 2, 3, 5, 6, 7, 8, 9, 10, 11, 13, 16, 17, 18, 19, 20, 21, 22, 23, 25, 26, 27, 29, 31, 32, 37, 38, 41, 45, 47, 50, 56, 63, 65, 72, 76, 77, 78, 80, 90, 91, 92, 93, 94, 100
crop insurance, 2, 3, 5, 6, 8, 9, 10, 11, 13, 17, 18, 19, 20, 21, 22, 23, 26, 27, 31, 32, 37, 38, 47, 50, 63, 65, 76, 77, 78, 80, 90, 92, 93, 94
crop production, 16
crop residue, 18
customers, 7, 38, 49
cyclones, 55

D

damages, vii, 15, 41, 58, 77, 82, 86, 88, 89, 95, 96, 97
data collection, 82
data set, 82, 84, 96, 97
database, 83, 95, 97
decision makers, 26, 79, 95
Department of Agriculture, 46, 82
Department of Commerce, 48, 56, 80, 97
Department of Energy, 42, 49, 80
Department of Homeland Security, 3, 4, 13, 28, 32, 46, 50, 83, 87, 99
Department of the Treasury, 28, 76, 89
Departments of Agriculture, 42, 48, 79
depth, 57, 80, 89
detection, 95
DHS, 3, 5, 28, 29, 30, 32, 48, 79, 87, 99
disaster, 4, 5, 7, 8, 9, 11, 13, 19, 25, 26, 34, 35, 37, 44, 50, 57, 64, 66, 67, 74, 86, 87, 88, 90, 91, 93, 94, 95, 96
disaster assistance, 4, 7, 8, 35, 44, 50, 64, 86, 87, 88, 90
disaster relief, 5, 9, 11, 19, 34, 35, 64, 74, 88
Disaster Relief Fund (DRF), 3, 5, 34, 36, 64
disclosure, 23, 24, 38
dissatisfaction, 90
distribution, 55, 64
District of Columbia, 82, 96
documentary evidence, 85
draft, 28, 42, 48, 56, 79, 80
drawing, 84
drought, viii, 4, 21, 41, 42, 45, 47, 51, 54, 55, 63, 65, 80, 96
drying, 55

E

earnings, 73
earthquakes, 13, 36, 59, 82
economic activity, 56, 66
economic consequences, 39
economic damage(s), 57, 63, 64
economic development, 66, 95
economic losses, 64, 66, 94
emergency, 34, 91
empirical studies, 31
energy, 53

enforcement, 26, 87
engineering, 88
environment, 37, 48, 72
erosion, 3, 17, 22, 27
Europe, 46, 84
evidence, 6, 30, 33, 53, 95
Executive Order, 10, 18, 19, 22, 27, 28, 29, 35, 36
expertise, 9, 46
exposure, vii, 1, 2, 3, 5, 6, 9, 10, 11, 13, 14, 16, 17, 19, 20, 21, 24, 25, 27, 30, 31, 34, 35, 39, 42, 47, 48, 49, 59, 67, 72, 73, 74, 75, 76, 77, 78, 79, 80, 88
extracts, 92
extreme cold, 56
extreme weather events, 2, 10, 21, 51, 57, 95

F

Fair Access to Insurance Requirement, 3, 7, 44, 50, 83
Farm Bill, 90
farm land, 38
farmers, 3, 4, 8, 9, 17, 18, 21, 22, 27, 29, 50, 63, 90, 91, 92, 93
FCIC, viii, 41, 42, 44, 45, 46, 48, 50, 59, 60, 63, 67, 68, 70, 72, 75, 76, 77, 78, 82, 83, 90, 91, 93, 94, 97, 99, 100
federal advisory, 97
federal agency, 34, 35, 39
federal assistance, 5
Federal Crop Insurance Corporation, viii, 41, 44, 45, 79, 82, 90
federal crop insurance program, vii, 1, 2, 4, 8, 11, 13, 22, 31, 90, 91, 92
Federal Emergency Management Agency (FEMA), vii, 1, 2, 3, 4, 5, 6, 8, 17, 18, 19, 20, 21, 22, 25, 26, 27, 28, 29, 30, 31, 32, 34, 35, 36, 37, 44, 46, 50, 57, 64, 76, 79, 82, 87, 88, 89, 99
federal funds, 76
federal government, 4, 5, 8, 9, 20, 21, 26, 27, 29, 34, 38, 39, 48, 50, 59, 67, 72, 78, 86, 87, 92, 94, 99
federal law, 3, 20, 22, 27, 29
financial, vii, 1, 4, 5, 7, 20, 21, 22, 35, 45, 48, 49, 50, 59, 67, 73, 75, 76, 77, 86, 90, 91, 96
financial condition, 5, 86
financial markets, 20
financial resources, 73
flexibility, 24
flooding, vii, 13, 17, 25, 35, 41, 42, 45, 47, 50, 51, 63, 86, 87, 89
floods, 4, 7, 26, 38, 45, 48, 50, 72, 77, 86, 88, 90, 94, 96
fluctuations, 35, 53, 90
force, 10, 18, 26, 36, 75, 93, 97, 100
formula, 64
fraud, 4
frost, 55, 63
funding, 19, 26, 34, 86
funds, 19, 20, 34, 37, 64, 75, 76, 89

G

GAO, vii, viii, 1, 2, 11, 12, 15, 33, 34, 35, 36, 37, 38, 41, 42, 43, 46, 48, 52, 60, 61, 62, 65, 68, 69, 70, 81, 84, 85, 97, 98, 99, 100
geography, 49
Germany, 66, 94, 98
Global Change Research Act, 97
governments, 26, 45
grants, 19
Great Depression, 50, 90
greenhouse, 51, 53, 94, 95
greenhouse gas(es), 51, 53, 94, 95
greenhouse gas emissions, 51, 95
gross domestic product, 82
growth, 2, 14, 42, 45, 47, 53, 59, 67, 72, 78
guidance, 18, 19, 22, 23, 28, 29, 37, 84
guidelines, 75
Gulf Coast, 14, 74

H

health, 7, 45, 78

height, 88
historical data, 36, 74, 85, 92
history, 17, 47, 53, 92, 94
homeowners, 25, 38, 63, 64, 87
homes, 21, 22, 25, 56, 67, 76, 87
Housing and Urban Development, 44, 86
HUD, 44, 86, 88
human, 7, 10, 45, 47, 53, 74, 94, 96
human activity, 45
Hurricane Andrew, 25, 38, 66
Hurricane Katrina, 5
hurricanes, 4, 16, 25, 26, 38, 39, 42, 47, 49, 55, 57, 59, 64, 65, 67, 74, 83, 84, 86, 88, 89, 96, 97

I

identification, 8
identity, 99
imbalances, 77
incidence, 47, 49
income, 38, 76, 88, 89, 90, 93, 96
income tax, 38
individuals, 5, 7, 9, 25, 35, 44, 45, 49, 64, 75, 79, 85
industries, 7, 49
industry, 2, 3, 6, 9, 14, 23, 24, 25, 31, 32, 35, 38, 39, 43, 44, 45, 46, 48, 56, 64, 67, 73, 74, 75, 78, 83, 84, 85
inflation, 10, 11, 14, 82
infrastructure, 10, 49
institutions, 35, 57
insurance policy, 3, 9, 21, 22, 44, 51, 88, 91
integration, 87
internal processes, 96
international standards, 39
investment(s), 10, 26, 38, 75
irrigation, 3, 22
issues, 5, 26, 49, 80, 94, 95

L

lakes, 54
landfills, 53
legislation, 5, 19, 35
lending, 35
levees, 86
light, 66, 78, 88
litigation, 84
livestock, 7, 45, 91
local government, 25, 34

M

magnitude, 4, 9, 15, 38, 42, 47, 51, 54, 72, 78
man, 10
management, 7, 8, 22, 28, 30, 38, 50, 73, 74, 84, 85, 87, 88, 89
mapping, 8, 17, 27, 30, 34, 35, 87
market share, 32, 84
materials, 66
measurement, 25
media, 98
median, 11, 35
methodology, 6, 17, 46, 92
Miami, 67
missions, 10, 18, 27, 78
Mississippi River, 14, 86
Missouri, 57
models, 15, 24, 38, 42, 45, 47, 48, 53, 54, 55, 73, 74, 88, 96, 97, 98
moisture, 54, 63

N

NAS, 44, 46, 51, 53, 64, 81, 97, 98
National Academy of Sciences, 33, 35, 44, 46, 81
National Climate Assessment, vii, 1, 4, 9, 16, 30, 33, 34, 36
National Flood Insurance Program, vii, 2, 3, 4, 31, 34, 36, 39, 41, 44, 45, 79, 82, 85, 100
National Research Council, 4, 9, 97
natural disaster(s), 4, 6, 32, 37, 45, 50, 64, 85, 90, 91, 95
natural resources, 10, 27

Natural Resources Conservation Service, 18
NOAA, 4, 14, 35, 44, 56, 64, 65, 74, 77, 84, 85
North America, 16, 47, 56
NRC, 4, 9, 35

O

oceans, 47, 53, 54
Office of Management and Budget, 64
officials, vii, 2, 3, 6, 17, 18, 19, 22, 25, 30, 32, 34, 35, 38, 39, 46, 48, 56, 72, 77, 78, 83, 84, 85
operating costs, 93
operations, 7, 18, 22, 45, 56, 75, 77
opportunities, 10, 26, 39
orbit, 47, 53
oversight, 50, 79, 87

P

participants, 33, 66, 76, 94, 99
peer review, 36, 94, 96
permafrost, 54
policy, 3, 8, 18, 21, 24, 30, 45, 64, 77, 81, 94
policymakers, 77, 78
pools, 5, 45, 50, 83, 97
population, 14, 31, 42, 47, 57, 59, 66, 67
population density, 14, 66
portfolio, 19, 73
potential benefits, 38
precipitation, 16, 51, 54, 55, 80
predictability, 35
preparedness, 26, 35, 38
President, 26, 97
price index, 82
price signals, 20, 21, 27
private insurers, vii, viii, 1, 3, 5, 6, 7, 9, 32, 41, 42, 45, 46, 47, 48, 50, 59, 72, 73, 74, 75, 77, 78, 79, 80, 85, 96
private property insurers, vii, 1
private sector, vii, 2, 4, 5, 6, 7, 10, 19, 25, 31, 34, 39, 45, 50, 64, 75, 76, 77, 84, 91
probability, 37, 56, 65
producers, 16, 29, 76, 90, 92, 93, 94
professionals, 83
profit, 43, 96
project, 45, 90
protection, 88, 90, 91, 92, 94
public health, 10
public insurers, vii, 1, 5, 6, 9, 21, 24, 25, 32
public officials, 82
public sector, vii, 1, 4, 7, 45, 48, 50, 74
publishing, 48
Puerto Rico, 82

Q

questionnaire, 84

R

rainfall, 54
rating agencies, 85
real estate, 37, 42, 47, 59, 66, 67
real property, 37, 82
realism, 96
recommendations, 5, 17, 26, 27, 28, 29, 34
recovery, 34, 74
recurrence, 88
reflectivity, 53
reform, 3, 10, 17, 27, 32, 34, 38, 86, 90, 99, 100
regulations, 8, 22, 25, 35, 50, 76, 87
regulatory affairs, 85
reimburse, 76
reinsurance, 3, 6, 19, 24, 25, 32, 34, 37, 39, 49, 58, 73, 74, 75, 77, 84, 85, 94, 97
reliability, 6, 31, 34, 38, 39, 49, 64, 82, 98
relief, 11, 19, 44
requirements, 8, 22, 27, 35, 86, 91
researchers, 33, 57, 84
reserves, 49, 76, 89
Residential, 39
resilience, 9, 10, 18, 22, 26, 27, 28, 29, 39
resources, 7, 27, 39, 50, 80, 99

response, viii, 5, 9, 10, 18, 19, 30, 41, 84, 86, 95, 97
restrictions, 7, 25, 99
revenue, 63, 76, 90, 91, 92, 93
risk assessment, 84, 85, 88, 92
risk management, 7, 17, 23, 24, 38, 42, 46, 48, 72, 73, 75, 77, 84, 85, 99
risks, 3, 4, 7, 9, 10, 17, 18, 19, 21, 22, 23, 24, 25, 26, 27, 30, 31, 37, 38, 39, 45, 48, 49, 50, 55, 56, 59, 65, 74, 75, 77, 78, 86
RMA, vii, 1, 2, 4, 6, 8, 16, 17, 18, 19, 20, 21, 22, 26, 28, 29, 32, 36, 37, 38, 44, 46, 50, 72, 80, 82, 84, 90, 91, 92, 93

S

safety, 10
SAP, 44, 96
scarcity, 18
science, 18, 26, 42, 47, 53, 54, 55, 96
scope, 6, 46, 64, 72, 80, 81, 82, 88
sea level, 9, 54
sea-level, 3, 7, 9, 17, 19, 22, 27, 30, 35
sea-level rise, 3, 7, 9, 17, 19, 22, 27, 30, 35
Secretary of Agriculture, 13, 28, 48, 79
Secretary of Homeland Security, 48, 79
security, 77
seed, 22
self-sufficiency, 76
Senate, 44, 99
sensitivity, 84
services, 97
shareholders, 75
shorelines, 86
shortfall, 63, 92
signals, 3, 20, 21, 27, 29
simulation(s), 78, 88
society, 73
soil erosion, 18
specialists, 85
specialty crop, 63, 91
spending, 9, 91
stability, 13
stakeholders, 30
state(s), 5, 6, 7, 8, 9, 14, 19, 25, 26, 29, 30, 32, 34, 37, 38, 39, 42, 45, 47, 50, 54, 55, 57, 67, 74, 75, 82, 83, 87, 93, 96, 97, 99, 100
state planning, 37
state regulators, 38
statistics, 85
statutes, 75, 76
statutory authority, 20, 86, 89
stock, 75
storms, 15, 26, 56, 57, 59, 65, 94, 96, 98
stress, 75
structure, 20, 37, 66, 67, 79, 97
Subsidies, 38
subsidy, 5, 21, 29, 38, 90
supervision, 13
supplier, 97
surplus, 76
sustainability, 99
synthesis, 44, 96

T

target, 63, 88, 92, 93
Task Force, 10, 18, 19, 22, 26, 27, 28, 30, 36, 39
taxpayers, 39
technical comments, 28, 49, 80
techniques, 31, 36, 88
technological advancement, 16
technological advances, 16
technology, 22, 26, 30
temperature, 16, 38, 42, 45, 47, 51, 53, 54, 55
tension, 22
terrorist acts, 82
testing, 82
theft, 86
thermal expansion, 54
threats, 77
time frame, 35
time periods, 15
time series, 18, 95
tornadoes, 96
transformation, 4

Treasury, 28
tropical storms, 56

U

U.S. Department of Agriculture (USDA), 4, 5, 8, 13, 18, 28, 29, 44, 48, 50, 79, 80, 90, 91, 93, 94
U.S. Treasury, 5, 20
underwriting, 7, 23, 24, 25, 49, 83
uninsured, vii, 1, 2, 5, 6, 10, 11, 14, 16, 25, 31, 42, 45, 46, 64, 80, 83, 86
universities, 9, 46, 85
updating, 34, 39

V

validation, 83
valuation, 97
variations, 96
vehicles, 56, 82
venue, 18
victims, 86, 87
volatility, 58
vulnerability, 3, 10, 22, 27, 29, 56, 80, 95

W

Washington, 33, 34, 35, 36, 37, 38, 81, 97, 98, 99, 100
waste, 4
water, 3, 18, 22, 27, 37, 47, 54
water resources, 18
wealth, 31
weather disasters, viii, 4, 41
weather patterns, 31, 45
weather related events, 94
weather-related losses, vii, viii, 1, 3, 4, 23, 24, 25, 41, 42, 45, 46, 47, 58, 59, 64, 66, 67
welfare, 72
White House, 19, 26, 37, 39
wind speed(s), 55
World Bank, 36
worldwide, 24

Y

yield, 63, 90, 92, 93, 94, 100